Marine Algae of

Northern Ireland

MARINE ALGAE OF
NORTHERN IRELAND

by
Osborne Morton

ULSTER
MUSEUM
Botanic Gardens, Belfast, BT9 5AB

Ulster Museum
Botanic Gardens, Belfast, BT9 5AB
1994

ISBN 0 900761 28 8
Ulster Museum Publication No. 271

Printed in Northern Ireland by
W&G Baird Ltd

Contents

Acknowledgements

I wish to express my sincere gratitude to all who assisted me in this work. I especially thank Mr Paul Hackney for suggesting this project and for his help, encouragement and advice on many occasions throughout, Dr Yvonne M. Chamberlain for reviewing the section on the Corallinales and Dr Christine Maggs for reviewing the entire manuscript and giving considerable help in a detailed and critical examination.

I am most indebted to all who collected, determined and confirmed specimens or provided other information - all are acknowledged in the text. However I especially wish to thank: the late Dr E.M. Burrows, Dr H. Blackler, Dr Y.M. Chamberlain, the late Professor P.S. Dixon, Dr M.D. Guiry, Mrs L.M. Irvine, Dr C.A. Maggs, and Dr G. Russell. Much of the typing onto word-processor was carried out by Mrs Irene McKeown. The work for this publication has been carried out during my employment by the Ulster Museum and I acknowledge the support of Dr David Erwin, Keeper of Botany and Zoology and Mr Philip Doughty, Head of the Division of Sciences whose help was most valuable. I am also most grateful to Mr John Nolan, the Director and to the Trustees of the Ulster Museum.

I am most grateful to the Ulster Museum Division of Sciences, the Esmé Mitchell Trust, the Friends of the Ulster Museum, the Department of the Environment (NI) Environment Service, the Belfast Natural History and Philosophical Society, the Belfast Naturalists' Field Club and the National Environment Research Council (recommended by The Linnean Society of London), for without their financial support this book could not have been published.

The maps are based on those of Paul Hackney and I am most grateful for his permission to use them. I am also indebted to Dr Julia D. Nunn, Mr Bernard E. Picton and Mr G. V. Day for permitting me to publish some of their photographs and to the Photographic Service of the Ulster Museum for photographing the illustration of W. Thompson, some pages from Templetons' *Hibernian Flora* and a number of algal specimens.

I express my thanks to Mr Roy Service and his Department of Design and Exhibition Services and especially to Mr James Hanna for the design and artwork necessary for publication. Finally and most importantly I would like to thank my wife, Ann, for her assistance in field work and valuable comments on the final draft of this book.

Introduction

Contents and Arrangement

This work is based on the style of *A Flora of the North-east of Ireland* by S.A. Stewart and T.H. Corry (1888) but deals with the marine algae which were not included in that work. It covers only Northern Ireland, i.e. the coasts of Counties Londonderry, Antrim and Down (Fig. 1), with the broad aim of providing a general list of sites from which the Chlorophyta, Phaeophyta and Rhodophyta have been recorded. It excludes the planktonic, unicellular (with a few exceptions) and freshwater species. A chapter by the author on the Characeae is included in the third edition of Stewart and Corry's *Flora of the North-east of Ireland*, compiled and edited by P. Hackney (1992).

This is not a publication dealing with the nomenclatural issues. The classification and nomenclature are based on South and Tittley (1986) but with alterations resulting from recent research and publications. The species and subspecific taxa are arranged alphabetically within each genus. The best known synonyms are listed but no attempt is made to provide a complete list of synonyms as this is beyond the scope of this work. Some English names are also noted.

The three counties from which the species have been recorded are noted using the initials **D** = Down, **A** = Antrim and **L** = Londonderry opposite the species name. After a short introductory note, the sites from which each species has been recorded are listed under the county name. The order in which they are listed is, in general, west to east on the northerly coasts of Counties Londonderry and Antrim and north to south on the east coasts of Counties Antrim and Down. A Topographical Index giving a four figure grid reference and detailing the location of these sites is included (q.v.). In most cases the dates have been obtained directly from the specimens or references. However, in some cases it has been necessary to estimate the date. Such estimates may be based on the date of publication of the record or from the date of botanical activity, or indeed death, of the collector. These appear as "fl", "c." or "pre-". For example: "pre-1825" is based on the death of the collector - John Templeton. Those species which are common and with too many records to be detailed are simply recorded: "Common", "Very common", "Frequent", "Widespread" or "Abundant" as appropriate. Each record is followed by an indication of its source. Those frequently used are reduced to abbreviations: a collector and /or determiner e.g. [OM], in square brackets, if supported by a voucher specimen usually in the Ulster Museum [**BEL**] or (OM), in round brackets, if not supported by a voucher specimen. A list of the abbreviations of collectors and determiners will be found in: "Collectors, Determiners, Recorders and Authors" q.v.

An un-published report or publication which is frequently referred to may also be abbreviated e.g. (*Phycol Brit*). A list of these abbreviations is included under: "References I. Abbreviated" q.v. A less frequently used reference, e.g. Morton, O. 1978, may be

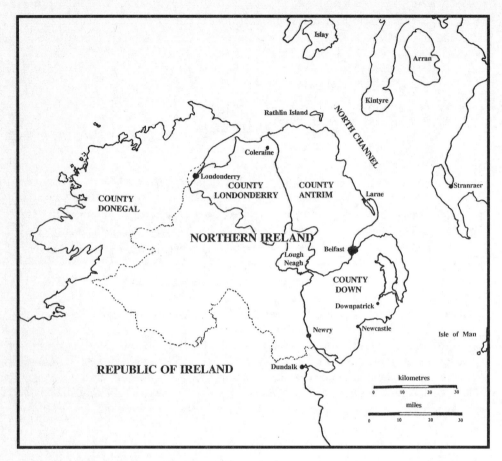

Fig. 1. The coastline of Northern Ireland in relation to the Republic of Ireland and Scotland. (Based on Hackney, 1992).

indicated as: (OM 1978). The initials of authors will also be found in the list of abbreviations: "Collectors, Determiners, Recorders, and Authors" q.v., and the full details of the reference under: References II. Other References". A rarely used reference is not abbreviated and is shown under the full name of the author and the date published.

A full record should therefore include: the site, date and source, along with an indication as to whether or not the record is supported by a voucher specimen. In some cases the the determiner of the specimen may also be noted .

Occasionally in the case of rare or interesting records a number preceded by the letter "F" is noted. This refers to the individual specimen preserved in the Ulster Museum Herbarium [**BEL**].

It has not been possible to confirm the determination of all the records. However, as discussed above, the source has been noted and some doubtful records have been omitted. Some genera are particularly difficult and further research is required on those such as *Ceramium* and *Audouinella*. Much work has recently been done on the Corallinaceae and Ceramiales and no doubt when published the nomenclature employed in this flora will in some cases be out of date. Certain subspecies, varieties, formae and ecads are also difficult and opinions differ concerning their taxonomic position. Further, due to the difficulty of identification, especially of the smaller algae, many are under-recorded.

Distribution

The coastlines of the counties of Northern Ireland vary from exposed to very sheltered and from short to relatively long, they are thus relatively rich and worthy of study. Our records go back to the latter years of the 18th Century when John Templeton (1766 - 1825) collected and recently there have been littoral and sublittoral surveys which have added considerably to our knowledge, especially that of the sublittoral flora.

Londonderry has a relatively short rocky coastline, much of it within the sheltered Lough Foyle, with only a short rocky shore on the exposed north coast around Portstewart. In general, marine algae require a firm surface on which to grow, and muddy or sandy shores have few if any seaweeds save those which are drift. County Antrim consists of the exposed north coast, including Rathlin Island, and the east coast, much of which is steep headlands and cliffs, with the exception of the sheltered Larne and Belfast Loughs. The County Down coastline is also long due to the Ards peninsula. It is relatively sheltered and includes the very sheltered Strangford and Carlingford Loughs. Due to its short coastline the records of algae from County Londonderry are relatively few and it is poorly represented compared to Counties Antrim and Down.

Altogether, records for 356 species of Chlorophyta, Phaeophyta and Rhodophyta have been assembled. There are, however, differences of opinion as to how certain entities should be treated, some workers considering certain populations as different species while others consider them to be conspecific. Where there is doubt, they have been recorded separately in this flora as it will be easier to unite them in a revision if and when the position is clarified than it will be to separate them once united. Further, for a number of species, the records are so few and so old that without voucher material, as is often the case, there is considerable doubt concerning the determinations. A recent example of this came to light concerning a freshwater alga, not included in this flora, collected in 1815 and identified then as *Thorea ramosissima* Bory. All records of this species in Ireland are based on a note by J. Templeton (John, Moore and Johnson, 1990). Fortunately voucher specimens had been preserved and in 1989, 174 years after they had been collected, they were redetermined as a species of *Batrachospermum* (John, Johnson and Moore, 1989). If voucher specimens had not been preserved this redetermination could never have taken place.

Of the 356 species, 77 are considered very rare, having only three or fewer citations. In a number of cases the species may be simply under-recorded due, amongst other reasons, to small size and/or difficulty of identification. Nevertheless 77 is a high proportion of the total. In some cases the records are old, some by almost two centuries and may be considered unreliable, others are not supported by voucher specimens. This should be borne in mind when reading the following discussion.

It has been suggested by Guiry (1978), amongst others, that the ratio of the numbers of species of Rhodophyta plus Chlorophyta to Phaeophyta (R + C: P) can be used to assess the affinities of marine algal floras in the north Atlantic. There is a general decrease in the ratio northwards as the red algae decrease in numbers and make up a smaller proportion of the total flora. Ratios less than 3 are taken to indicate a cold water flora and ratios of 6 or more indicate a tropical flora (Guiry, 1978). The ratio for the north-east of Ireland is 2.83 (Table 1) and includes all species in this flora. If doubtful or rare records are excluded, the ratio is a similar one of 2.9. This is a relatively high figure - higher than expected and higher than that for Ireland in Guiry (1978).

Table 1

Floristic ratios for the Rhodophyta (R), Chlorophyta (C) and Phaeophyta

Place	R	C	P	(R+C)/P

A. Northern Ireland based on this flora:-

Place	R	C	P	(R+C)/P
	198	65	93	2.83

B. Other regions of the north Atlantic (Earll and Farnham, 1983):-

Place	R	C	P	(R+C)/P	Reference:-
Newfoundland	55	23	41	1.9	Hooper *et al.* (1980)
Greenland	38	36	68	1.1	Lund (1959)
Iceland	83	54	72	1.9	Caram & Johnson (1972)
Faeröes	83	44	73	1.7	Børgesen (1905)
Helgoland (Germany)	73	53	56	2.2	Kornmann & Sahling (1977)
Netherlands	62	58	54	2.2	Den Hartog (1959)
British Isles	334	115	207	2.2	Parke & Dixon (1976)
Ireland	246	87	147	2.3	Guiry (1978)
Roscoff (N W France)	286	96	160	2.4	Feldmann (1954); Feldmann & Magne (1964)
Portugal	246	60	98	3.1	Ardré (1970)
W. Mediterranean	175	40	65	3.3	Coppejans (1980)
Albères (Med.)	259	79	88	3.8	Feldmann (1937)
Canaries	202	69	55	4.9	Børgesen, in Feldman (1937)
Trinidad	71	38	24	4.5	Richardson (1975)
E. Africa	327	143	97	4.9	G. Lawson (unpublished)
Texas	53	21	14	5.3	Edwards (1970)
Jamaica	161	123	47	6.0	Chapman (1961, 1963)

Over 70% of Irish algae have been recorded in Northern Ireland and nearly 60% of the
species recorded from the British Isles have been recorded from the three counties of
Northern Ireland. This is detailed in Table 2 below.

Table 2

Total number of algal species recorded from Northern Ireland, all Ireland and the British Isles
(Including some doubtful records)

	Chlorophyta	Phaeophyta	Rhodophyta	Totals
Northern Ireland	65	93	198	356
All Ireland*	83	143	267	493
% of Irish species recorded in Northern Ireland	78.3%	65%	74.2%	72.2%
Total British Isles species * (excl. Shetland Islands)	103	192	319	614
% of British Isles species recorded in Northern Ireland	63.1%	48.4%	62.1%	58%

* Based on South and Tittley (1986).

Of the 356 species recorded in this flora, 20 are not recorded from Ireland in South and
Tittley (1986), although in some cases they are united in synonymy under another name. Of
the remaining 336, over half (189) are also recorded on the Atlantic coast of North
America. Of these, 47 are Chlorophyta, 55 are Phaeophyta and 87 are Rhodophyta. In some
cases these species are recorded all along the north Atlantic coasts from Portugal via
Spitsbergen, Iceland, Greenland to the east coast of North America from arctic eastern
Canada to Virginia in the U.S.A. (South and Tittley, 1986).

306 species of algae recorded in this flora are also found in Scotland (excluding the
Shetland Islands). Of these 66 are apparently southern species being found no further north
than Scotland (excluding the Shetland Islands) and are therefore nearing their northern limit
(Table 3). In winter, the sea temperatures on the west coast of Ireland are higher than on the
east (Lewis, 1964). This no doubt explains the presence of some species further north on
the west coast of Ireland than on the east and not found on the north-east coast. Table 4
gives a list of Irish algae not yet recorded from Northern Ireland.

Table 3

Some species considered to be near or at their northern limit in Northern Ireland
(Based on South and Tittley, 1986)

Chlorophyta

Urospora bangioides (Harv.) Holm. et
    Batt.
Enteromorpha ralfsii Harv.
Chaetomorpha crassa (C. Ag.) Kütz.
Cladophora pellucida (Huds.) Kütz.
C. vagabunda (L.) Hoek
Codium adhaerens C. Ag.
C. bursa (Olivi) C. Ag.
C. tomentosum Stackh.

Phaeophyta

Petrospongium berkeleyi (Grev.)
    Näg.
Sphacelaria bipennata (Kütz.) Sauv.
S. fusca (Huds.) S.F. Gray
Halopteris filicina (Grat.) Kütz.
Dictyopteris membranacea (Stackh.) Batt.
Taonia atomaria (Woodw.) J. Ag.
Carpomitra costata (Stackh.) Batt.
Desmarestia dresnayi Lamour. ex Leman
Bifurcaria bifurcata Ross
Cystoseira baccata (Gmel.) Silva
C. nodicaulis (With.) Roberts
C. tamariscifolia (Huds.) Papenf.

Rhodophyta

Audouinella caespitosa (J. Ag.) Dixon
A. concrescens (Drew.) Dixon
Schmitziella endophloea Born. et Batt.
Atractophora hypnoides P. et H. Crouan
Naccaria wiggii (Turn.) Endl.
Gelidiella calcicola Maggs et Guiry
Halymenia latifolia P. et H. Crouan
Kallymenia microphylla J. Ag.
Peyssonnelia harveyana J. Ag.
Corallina elongata Ellis et Sol.
Haliptilon virgatum (Zanard.) Garbary et
    Johansen
Mesophyllum lichenoides (Ellis) Lemoine
Pneophyllum concollum Chamberlain

P. microsporum (Rosenv.) Chamberlain
Schmitzia hiscockiana Maggs et Guiry
S. neapolitana (Berth.) Lagerh. et Silva
Gymnogongrus crenulatus (Turn.) J. Ag.
G. griffithsiae (Turn.) Martius
Schottera nicaeensis (Duby) Guiry et
    Hollenberg
Gigartina acicularis (Roth) Lamour.
Sphaerococcus coronopifolius Stackh.
Calliblepharis ciliata (Huds.) Kütz.
C. jubata (Good. et Woodw.) Kütz.
Rhodymenia delicatula P. Dang.
R. holmesii Ardiss.
R. pseudopalmata (Lamour.) Silva
Gastroclonium ovatum (Huds.) Papenf.
Antithamnionella spirographidis
    (Schiffner) Wollas.
Callithamnion granulatum (Ducluz.)
    C. Ag.
C. tetricum (Dillw.) S.F. Gray
Ceramium graditanum (Clem.) Cremades
Halurus equisetifolius (Lightf.) Kütz.
Ptilothamnion pluma (Dillw.) Thur.
Spermothamnion mesocarpum (Harv.)
    Chemin.
Spyridia filamentosa (Wulf.) Harv.
Gonimophyllum buffamii Batt.
Drachiella heterocarpum (Chauv. ex
    Duby) Maggs et Hommersand
Radicilingula thysanorhizans (Holm.)
    Papenf.
Bostrychia scorpioides (Huds.) Mont.
Chondria dasyphylla (Woodw.) C. Ag.
Laurencia obtusa (Huds.) Lamour.
Polysiphonia atlantica Kapraun et J. Norris
P. denudata (Dillw.) Harv.
P. furcellata (C. Ag.) Harv.
P. subulifera (C. Ag.) Harv.
Boergensiella thuyoides (Harv.)
    Kylin

Table 4

Some Irish algae not yet recorded from the coasts of Northern Ireland

(Based on South and Tittley, 1986)

Chlorophyta
Uronema? curvata Printz
Eugomontia sacculata Kornm.
Pseudendoclonium fucicola
    (Rosenv.) Nielsen
P. submarinum Wille
Acrochaete repens Pringsh.
Ochlochaete hystrix Thwaites
Pseudopringsheimia confluens
    (Rosenv.) Wille
Cladophora aegagropila (L.) Rabenh.
C. battersii Hoek
C. coelothrix Kütz.
C. dalmatica Kütz.
C. prolifera (Roth) Kütz.
Ostreobium quekettii Born. et Flah.

Phaeophyta
Dichosporangium chordariae Wollny
Feldmannia caespitula (J. Ag.)
    Knoepffler-Peguy
F. simplex (P. et H. Crouan) Hamel
Giffordia mitchelliae (Harv.) Hamel
Gononema aecidioides (Rosenv.) Pedersen
Herponema solitarium (Sauv.) Hamel
Kuetzingiella battersii (Born.) Kornm.
K. holmesii (Batt.) Russell
Mikrosyphar polysiphoniae Kuck.
M. porphyrae Kuck.
Streblonema breve (Sauv.) De Toni
S. fasciculatum Thur.
S. parasiticum (Sauv.) Levring
S.?sphaericum (Derbès. et Solier) Thur.
S. stilophorae (P. et H. Crouan) Hamel
S. tenuissimum Hauck
S. zanardinii (P. et H. Crouan) De Toni
Waerniella lucifuga (Kuck.) Kylin
Petroderma maculiforme (Wollny) Kuck.
Pseudolithoderma roscoffensis Lois.
Sorocarpus micromorus (Bory) Silva
Chilionema foecundum (Strömfelt) Fletcher

C. ocellatum (Kütz.) Kuck.
C. reptans (P. et H. Crouan) Sauv.
Compsonema microspongium (Batt.) Kuck.
C. saxicola (Kuck.) Kuck.
Microspongium globosum Reinke
Myrionema corunnae Sauv.
M. magnusii (Sauv.) Lois.
M. papillosum Sauv.
Protectocarpus speciosus (Børg.) Kuck.
Corynophlaea crispa (Harv.) Kuck.
Cylindrocarpus microscopicus
    P. et H. Crouan
Microcoryne ocellata Strömfelt
Myriactula clandestina
    (P. et H. Crouan) Feldm.
M. haydenii (Gatty) Levring
M. rivulariae (Suhr) Feldm.
M. stellulata (Harv.) Levring
Stilopsis lejolisii (Thur.) Kuck.
Cladosiphon contortus (Thur.) Kylin
C. zosterae (J. Ag.) Kylin
Mesogloia lanosa P. et H. Crouan
Sauvageaugloia griffithsiana (Harv.) Kylin
Haplospora globosa Kjellm.
Sphacelaria caespitula Lyngb.
S. nana Kütz.
S. plumigera Holm.
S. plumula Holm.
S. rigidula Kütz.
Padina pavonia (L.) Lamour.
Leblondiella densa (Batt.) Hamel
Stictyosiphon soriferus (Reinke) Rosenv.
S. tortilis (Rupr.) Reinke
Giraudia sphacelarioides Derbès et Solier
Desmotrichum undulatum (J. Ag.) Reinke
Fucus distichus L.
Cystoseira foeniculaceus (L.) Grev.

Rhodophyta
Chroodactylon ornatum (C. Ag.) Basson
Colacodictyon reticulatum (Batt.) Feldm.

Neevea repens Batt.
Erythrotrichia welwitschii (Rupr.) Batt.
Porphyra amethystea Kütz.
Audouinella bonnemaisoniae (Batt.) Dixon
A. chylocladiae (Batt.) Dixon
A. ? corymbifera (Thur.) Dixon
A. endophytica (Batt.) Dixon
A. endozoica (Darb.) Dixon
A. ? parvula (Kylin) Dixon
A. seiriolana (Gibs.) Dixon
Meiodiscus spetsbergensis (Kjellm.)
    Saunders et McLachlan
Helminthocladia calvadosii (Duby) Setch.
Asparagopsis armata Harv.
Halosacciocolax kjellmanii Lund
Rhodophysema feldmannii Cabioch
R. georgii Batt.
Holmsella pachyderma (Reinsch) Sturch
Cryptonemia hibernica Guiry et Irvine
Dermocorynus montagnei P. et H. Crouan
Grateloupia filicina (Lamour.) C. Ag.
Peyssonnelia atropurpurea P. et H. Crouan
Choreonema thuretii (Born.) Schmitz
Clathromorphum circumscriptum
    (Strömfelt) Fosl.
Dermatolithon confinis (P. et H. Crouan)
    Boudouresque, Perret-Boudouresque
    et Knoepffler-Peguy
D. haplidioides (P. et H. Crouan) Fosl.
D. litorale (Suneson) Lemoine
Haliptilon squamatum (L.) Johanson,
    Irvine et Webster

Jania corniculata (L.) Lamour.
Leptophytum laeve (Strömfelt) Adey
Lithophyllum dentatum (Kütz.) Fosl.
Lithothamnion norvegicum (Aresch.)
    Kjellm.
Pneophyllum limatum (Fosl.) Chamberlain
P. myriocarpum (P. et H. Crouan)
    Chamberlain
P. rosanoffii Chamberlain
P. zonale (P. et H. Crouan) Chamberlain
Spongites fruticulosa Kütz.
Platoma marginifera (J. Ag.) Batt.
Gracilaria compressa (C. Ag.) Grev.
G. multipartita (Clem.) Harv.
Ceratocolax hartzii Rosenv.
Phyllophora sicula (Kütz.) Guiry et Irvine
Gigartina pistillata (Gmel.) Stackh.
G. teedii (Roth) Lamour.
Cruoria cruoriaeformis (P. et H. Crouan)
    Denizot
Gastroclonium reflexum (Zanard.) Kütz.
Ceramium flaccidum (Kütz.) Ardiss.
Crouania attenuata (C. Ag.) J. Ag.
Seirospora interrupta (Sm.) Schmitz
Pleonosporium borreri (Sm.) Näg.
Dasya corymbifera J. Ag.
Chondria capillaris (Huds.) M. Wynne
Halopitys incurvus (Huds.) Batt.
Lophosiphonia reptabunda (Suhr) Kylin
Polysiphonia foetidissima Cocks
P. simulans Harv.
Pterosiphonia pennata (C. Ag.) Falkenb.

In Northern Ireland, other algae are near their southern limit. The best known of these is *Odonthalia dentata* (L.) Lyngb. which is found on west Atlantic shores from west Greenland south to the maritime provinces of Canada (South and Tittley,1986). In Ireland it is rarely found south of County Down. It has been recorded from the Counties Donegal, Leitrim, Sligo and Mayo on the north and west coasts of Ireland and from Dublin on the east coast - but no further south (Guiry, 1978). The Isle of Man is its southern limit where it is "abundant in the subtidal" (J. M. Jones - pers. comm.). Other algae nearing their southern limit include: *Sphacelaria plumosa* Holm. (South and Tittley, 1986), *Lithothamnium glaciale* Kjellm. and *Ptilota gunneri* Silva, Maggs et Irvine. (Hardy and Aspinall, 1988). *P. gunneri* is reported along the west and south coast of Ireland to Waterford (Guiry, 1978), Isle of Man, Anglesey (Norton, 1985) and on the east coast of England no further south than Northumberland (Hardey and Aspinall, 1988). In County Down *Hildenbrandia crouanii* J. Ag. may be an under-recorded northern species, while *H. rubra* (Sommerf.) Menegh. - a closely related species - is more common in Northern Ireland and penetrates further south to France, north Spain and Portugal as well as along the Atlantic coast of north America (South and Tittley, 1986). *Haemescharia hennedyi* (Harv.) Vinogradova is also apparently near its southern limit. In Ireland *H. hennedyi* has only been reported from Counties Mayo, Cork and Down but in its sterile state it is not easily distinguishable from what used to be referred to as *Petrocelis cruenta* J. Ag., now known to be the tetrasporic phase of the common *Mastocarpus stellatus* (Stackh. in With.) Guiry.

The useful checklist, South and Tittley (1986), while it distinguishes records from Ireland and Scotland, unfortunately does not distinguish those of Wales from England. However of the 347 species listed in South and Tittley, and recorded in this flora, 338 are also recorded in England and Wales. This small difference of 9 is considered to be insignificant.

The distribution of some species is changing and amongst the better known examples is *Sargassum muticum* (Yendo) Fensholt - a Japanese species first recorded on the Isle of Wight in 1973. It is now spreading along the south coast of England and from Brittany to the Netherlands (Norton, 1985) and one day may reach Northern Ireland. *Colpomenia peregrina* Sauv. was first recorded in England in 1908 (Cotton, 1908) and in Northern Ireland in 1934 (Lynn, 1935). Since then it has been recorded in all three coastal counties of Northern Ireland, most coastal counties of Ireland (Guiry, 1978) and from Scotland (Maggs, 1986). *Trailliella intricata* Batt. was first recorded in Northern Ireland in 1972 (Morton, 1974) and was then thought to be a "true" species. However it was later discovered to be the tetrasporic phase of *Bonnemaisonia hamifera* Hariot, the gametangial phase of which has not yet been found in Northern Ireland but has been recorded from England since the last century (1893) and the west of Ireland since 1910 (Cotton, 1912). It may reach the north-east before long. *Gelidiella calcicola* Maggs et Guiry was recorded in Northern Ireland in 1983. Since then there have been further records of it from elsewhere in Ireland, Britain and France (Maggs and Guiry, 1987). The distribution of other species is no doubt also changing and efforts are made to record their distribution and to ensure changes do not pass unnoticed.

Algal phytogeography has received much recent coverage. One suggestion is that there are two floristic discontinuities along the European Atlantic coasts: one from north-west Brittany via Clare Island in County Mayo to the Faeröes, and the second from Arctic Europe to Spitsbergen. Both discontinuities show the disappearance of a large number of species. 93% of the species in the Faeröes also occur in north-west Brittany but only 37% of the species in north-west Brittany occur also in the Faeröes. The obvious decrease in the number of eulittoral species between Arctic Europe and Spitsbergen is considered to be due to the heavy scouring action of ice (van den Hoek and Donze, 1967). Maggs (1986) accepted the phytogeographic regions separated by floristic discontinuities in the north-east Atlantic by Hoek (1975) and noted the three regions: the warm temperate Mediterranean-Atlantic, the cold temperate Atlantic-Boreal and the Arctic. The warm temperate Atlantic region with its northern boundary at the 15°C summer isotherm, and the southern boundary at the 20°C winter isotherm, is itself divided into three provinces: Canaries, Mediterranean and Lusitanian (Maggs, 1986). The cold temperate Atlantic-Boreal has its southern boundary at the 15°C summer (10°C winter) isotherms and the northern boundary at the 10°C summer isotherm. This places northern Ireland and northern Britain in the cold temperate Atlantic-Boreal region.

Using species lists there appears to be a gradual change in the algal flora from the west of Ireland to the Orkneys. However analysis of species lists may conceal true phytogeographic affinities. This could be due to certain rare or difficult species not being recorded (Maggs, 1986) and it is clear that collecting from the sublittoral shows many species to have a wider distribution than the littoral records indicate. It is interesting at this point to comment that some of the old historic specimens may have been drift specimens from the sublittoral and in some cases may suggest a more accurate distribution for some species.

As has already been commented, the distribution of all species of algae found in Northern Ireland extends well beyond the British Isles and many are found on the east coast of America. Other species have invaded the country from elsewhere probably as a result of shipping. Some of our species are southern species approaching their northern limit of range while others are northern species approaching the southern limit of their range. Recent sublittoral collecting has revealed an extension to the range of many species and shown many to be more abundant than previously realised.

This flora records those species present in Northern Ireland. However, other small or difficult species, including parasitic species, may be under-recorded. These gaps in our knowledge are clear and this flora shows where further work is required.

Historical Background

The scientific study of marine algae of the north-east of Ireland began at the end of the 18th century with the work of John Templeton (1766 - 1825). The first edition of Stewart and Corry's *Flora...*(1888) and the second edition by Praeger (1938), although not including the marine algae, gave an account of his life and work. In the second edition R.Ll. Praeger also gave an account of the unpublished manuscripts of Templeton. Other publications including Kertland (1966) also record his work. Kertland (1967a) and Pilcher (1967) detail the algal specimens collected by, or associated with, Templeton which were then stored in the Queen's University Herbarium (**BFT**) and the Ulster Museum Herbarium (**BEL**). These were amalgamated in 1968 and are housed in the Ulster Museum.

Although by modern standards the combined algal collections of Templeton, just over 160 specimens, are very small, they are of considerable interest. They are mainly from Counties Down and Antrim; it seems he did not travel far and found enough to study in these counties. A few specimens collected or determined by other botanists: Robert Brown (1773 - 1858), Dawson Turner (1775 - 1858) and James Lawson Drummond (1783 - 1853), have also been found in the Templeton collections. The earliest alga, indeed the earliest botanical specimen, in the Ulster Museum is in the John Templeton collections and is of *Dumontia contorta* (S. Gmelin) Rupr. dated April 1798 (**BEL**: F208). It was recorded as: "Ulva filiformis of Hudson, D. Turner Found on the shore of Belfast Lough adhering to a mussel shell". Templeton, of course, also collected and studied flowering plants, ferns, mosses, liverworts and animals.

After Templeton, other botanists such as G.C. Hyndman (1796 - 1867), continued the collecting and studying of algae in the north-east of Ireland. However it was William Thompson (1805 - 1852) who, although primarily a zoologist, carried forward the work. By then (1830 - 1850's) there were other Irish biologists with whom he could communicate such as George Dickie (1812 - 1882), Rev. John Hutton Pollexfen (1813 - 1899), William Henry Harvey (1811 - 1866), William McCalla (Macalla) (c.1814 - 1849), Mrs Griffiths (1768 - 1858) and Miss Anne Elizabeth Ball (1808 - 1872) to name but a few. These names, with others, are to be found on specimens in Thompson's albums and it was thanks to botanists such as these, who exchanged and donated specimens, that knowledge of our algal flora has accumulated. Unfortunately Thompson died a young man at the age of only forty-seven. However between them Templeton and Thompson were: "the most noticeable of naturalists which Belfast has produced" (Praeger, 1949). In the Ulster Museum the algae collected by Thompson are mounted in five large volumes, one of foreign species, totalling 985 specimens, as well as many others in the main herbarium. There is also a collection of Thompson's in the National Botanic Gardens Herbarium, Glasnevin, Dublin (**DBN**).

The results of the work of these botanists, along with that of many others from elsewhere in Ireland, Britain and further afield were included and acknowledged by Harvey in: *A Manual of the British Algae...* ,(1841) and later in his: *Phycologia Britannica...,*(1846 - 1851) published with beautiful colour plates. To this day it is recognised as a valuable publication. The list of collectors, at this time, is quite long and indicates how much the interest in botany had blossomed. Many also collected flowering plants, ferns, lichens and mosses and they exchanged and donated specimens freely between each other.

Following Thompson, the next most important collector from Northern Ireland is William Sawers (fl. 1852 - 1856); relatively little is known of him (Kertland, 1967b). His specimens in the Ulster Museum are mostly from Counties Donegal, Londonderry and Antrim and all were collected between 1853 and 1856 (McMillan and Morton, 1979), the most interesting being *Desmarestia dresnayi* Lamour. ex Leman (F7645) noted by Sawers as: "A number of specs. got floating near Greencastle and Moville, mouth of Lough Foyle between Augt & December W Sawers *Desmarestia herbacea* Autumn 1853". A.D. Cotton saw this specimen in 1914 and thought the determination to be "uncertain" but "most likely" correct. South and Tittley (1986) included the name as a synonym under *D. ligulata* while Fletcher (1987) retained the species, as *D. dresnayi*, "pending further studies." We have only one other specimen (F5037) which, unlike the Sawers specimen, was collected attached during a dive by C.A. Maggs at Altacorry Head on Rathlin Island on 3 August 1985.

There seems to have been little phycological work in the latter half of the 19th century except for Henry Hanna (fl. 1898 - 1899), who is not mentioned in Britten and Boulger (1931) or Desmond (1977). Hanna's specimens collected during 1898 and 1899 are nicely pressed and are preserved in the Ulster Museum. He collected in Counties Mayo, Galway, Donegal, Antrim and Down and his specimens are well dated so that one can see he collected at the Gobbins in County Antrim on the 4th July 1898, at Torr Head in County Antrim on the 3rd, 4th and 5th August 1898, at Roundstone in County Galway in April 1899 to note but a few of the sites he visited. Although most are not signed or initalled by Hanna they show his handwriting and "H.H" has been added to them by a later worker. Hanna also published some papers in *The Irish Naturalist* and contributed a short section on algae to Bigger, Praeger and Vinycomb (1902).

Sylvanus Wear (1858 - 1902) was born in Northumberland but settled in Belfast in 1904 and collected algae, as well as other plants, between 1915 and 1917 in Counties Down, Antrim and Londonderry. His algal collection in the Ulster Museum is preserved as one entity, Collection 12, containing 273 specimens (Morton, 1977). He compiled the *Second Supplement* to Stewart and Corry's *Flora* in 1923 and a photograph of him will be found facing the title page of that publication.

At about the same time Margaret Williamson Rea (1875 - ?) collected in Counties Antrim and Down. She worked in the algal herbarium of Queen's University in about 1915 - 1916 arranging the specimens and was an Honorary Secretary of the Belfast Naturalists' Field Club with J.K. Charlesworth between 1916 and 1919. In 1934 she published a paper on W. Thompson's algal collection.

Margaret Constance Helen Blackler (1902 - 1981), was born, graduated and died in

Britain. She had an interest in algal ecology and her name is especially linked with *Colpomenia peregrina* (Sauv.) Hamel. She published several papers concerning this species (as *C. sinuosa*) between 1937 and 1981 at least two of which were about the species in Ireland. She also confirmed the determination of a specimen which I collected in 1976 from Ardkeen (F1682). She was co-author with N.F. McMillan (1953) of a paper on mollusca and algae from Port Ballintrae and showed a special interest in the algae of Lough Foyle (Irvine and Russell, 1982).

During the first half of this century, in spite of two world wars, other workers continued the study of algae. M.P.H. Kertland (1901 - 1991) worked as the curator of the herbarium of Queen's University and, as already noted, published papers concerning it. Her determinations and notes are to be found on many herbarium specimens written in blue ball-point pen. Other workers include Rev. William Rutledge Megaw (1885 - 1953) and N.F. McMillan (fl. 1950's).

In the second half of the 20th century work on the algae increased, inspired perhaps by Parke's *Preliminary Checklist...* in 1953 and its subsequent revisions (Parke and Dixon, 1964, 1968 and 1976). The frequency of these, together with the key to the genera by Jones (1962) itself reprinted with revisions in 1964 and 1968, indicates the interest in algae at this time. However, as Jones' key was only at generic level and Newton's (1931) *Handbook...* was well out of date, the British Phycological Society in 1953 undertook the work of writing an up to date British marine algal flora with keys and descriptions. To date volumes 1 (parts 1, 2A, 2B and 3A), 2, 3 (part 1) and 4 have been published. Volume 2 was written by Elsie May Burrows (1913 - 1983), who determined or confirmed many algae which I collected in Northern Ireland and are now stored in the Ulster Museum. The volume was posthumously published in 1991. In the same period a mapping scheme was started, to which I contributed, and a *Provisional Atlas* was published in 1985 (Norton, 1985).

Between 1982 and 1985 the Ulster Museum, in contract with the Department of the Environment (NI), carried out a survey of the animals and plants of the sublittoral in Northern Ireland. This provided considerable lists of algae and added over one thousand algal specimens from Northern Ireland, all well determined and detailed, to the Ulster Museum Herbarium as Collection 30. As the survey was sublittoral some of the specimens are of considerable interest and value as they had rarely, if ever before, been recorded from Northern Ireland. Some were species new to the Ulster Museum Herbarium.

Between 1984 and 1987 another survey, this time of the littoral, provided many littoral records as well as specimens which are now added, as Collection 37, to the Ulster Museum Herbarium. The encrusting algae had, until I took an interest in the group in 1976, been somewhat ignored, probably because of difficulty in collection and identification. Specimens of these have now been added to the herbarium.

Although seaweeds have now been studied for more than two hundred years, their taxonomy, classification and distribution are still in some cases uncertain. The aim of this publication is to bring together all available records of marine algae from Northern Ireland as Stewart and Corry did for the higher plants from the north-east over one hundred years ago in 1888.

References

Bigger, F.J., Praeger, R.Ll. and Vinycomb, J. 1902. *A Guide to Belfast and the Counties of Down and Antrim. Prepared for the meeting of the British Association by the Belfast Naturalists' Field Club.* M'Caw, Stevenson and Orr, Belfast.

Blackler, (M.C.) H. and McMillan, N.F. 1953. Some interesting marine mollusca and algae from Port Ballintrae, in north Ireland. *Ir. Nat. J.* **11**: 76.

Britten, J. and Boulger, G.S. 1931. *A Biographical Index of Deceased British and Irish Botanists.* Second edition. Revised and completed by A.B. Rendle. London.

Burrows, E.M. 1991. *Seaweeds of the British Isles.* **2** *Chlorophyta.* Natural History Museum, London.

Cotton, A.D. 1908. The appearance of *Colpomenia sinuosa* in Britain. *Bull. misc. Inf. R. bot. Gdns Kew.* **1908**: 73 - 77.

Cotton, A.D. 1912. Marine Algae, *In*: Praeger, R.Ll. A biological survey of Clare Island; in the County of Mayo, Ireland and the adjoining district. *Proc. R. Ir. Acad.* **31** sect. 1 (15): 1 - 178.

Desmond, R. 1977. *Dictionary of British and Irish Botanists and Horticulturists.* Taylor & Francis Ltd., London.

Earll, R. and Farnham, W. F. 1983. Biogeography. *In*: *Sublittoral Ecology. The Ecology of the Shallow Sublittoral Benthos.* (Eds Earll, R. and Erwin, D.G.). pp. 165 - 208. Clarendon Press, Oxford.

Fletcher, R. 1987. *Seaweeds of the British Isles.* **3** pt.1. *Fucophyceae* (*Phaeophyceae*) . British Museum (Natural History), London.

Guiry, M.D. 1978. A concensus and bibliography of Irish seaweeds. *Bibliotheca Phycologia* **44**: 1- 287.

Hackney, P. (Ed.) 1992. *Stewart & Corry's Flora of the North-east of Ireland.* Third edition. Institute of Irish Studies, The Queen's University of Belfast.

Hardy, F.G. and Aspinall, R.J. 1988. *An Atlas of the Seaweeds of Northumberland and Durham.* Northumberland Biological Records Centre, The Hancock Museum, The University of Newcastle upon Tyne. Special Publication No.3.

Harvey, W.H. 1841. *A Manual of the British Algae:...* John van Voorst, London.

Harvey, W.H. 1846 - 1851. *Phycologia Britannica...* London.

Hoek, C. van den, 1975. Photogeographic provinces along the coasts of the northern Atlantic Ocean. *Phycologia* **14**: 317 - 330.

Hoek, C. van den and Donze, M. 1967. Algal phytogeography of the European Atlantic coasts. *Blumea* **15**: 63 - 89.

Irvine, D.E.G. and Russell, G. 1982. Obituary. Margaret Constance Helen Blackler (1902 - 1981). *Br. phycol. J.* **17**: 343 - 346.

John, D.M., Johnson, L.R. and Moore, J.A. 1989. The red alga *Thorea ramosissima*: its distribution and status in the Thames catchment. *The London Naturalist* **68**: 49 - 53.

John, D.M., Moore, J.A. and Johnson, L.R. 1990. The red alga *Thorea* in the British Isles. *Br. phycol. Soc. Newsletter* No. 28: 11 - 12.

Jones, W.E. 1962. A key to the genera of British seaweeds. *Field Studies.* **1**(4): 1 - 32. Reprinted with revisions: 1964 and 1968.

Kertland, M.P.H. 1966. Bi-centenary of the birth of John Templeton, A.L.S. 1766 - 1825. *Ir. Nat. J.* **15**: 229 - 231.

Kertland, M.P.H. 1967a. The specimens of Templeton's algae in the Queen's University Herbarium. *Ir. Nat. J.* **15**: 318 - 322.

Kertland, M.P.H. 1967b. Some early algal collections in the Queen's University Herbarium. *Ir. Nat. J.* **15**: 346 - 349.

Lewis, J.R. 1964. *The Ecology of Rocky Shores.* English Universities Press, London.

Lynn, M.J. 1935. Rare algae from Strangford Lough. - Part 2. *Ir. Nat. J.* **5**: 275 - 283.

Maggs, C.A. 1986. Scottish marine macroalgae: a distributional checklist, biogeographical analysis and literature abstract - *Report for the Nature Conservancy Council, Peterborough* **635**: 1 - 137.

Maggs, C.A. and Guiry, M.D. 1987. *Gelidiella calcicola* sp. nov. (Rhodophyta) from the British Isles and northern France. *Br. phycol. J.* **22**: 417 - 434.

McMillan, N.F. and Morton, O. 1979. A Victorian album of algae from the north of Ireland with specimens collected by William Sawers. *Ir. Nat. J.* **19**: 384 - 387.

Morton, O. 1974. Marine algae of Sandeel Bay, Co. Down. *Ir. Nat. J.* **18**: 32 - 35.

Morton, O. 1977. Sylvanus Wear's algal collection in the Ulster Museum. *Ir. Nat. J.* **19**: 92 - 93.

Newton, L. 1931. *A Handbook of the British Seaweeds.* British Museum (Natural History), London.

Norton, T.A. (Ed.) 1985. *Provisional Altas of the Marine Algae of Britain and Ireland.* Biological Records Centre, Institute of Terrestial Ecology and Nature Conservancy Council.

Parke, M. 1953. A preliminary check-list of British marine algae. *J. mar. biol. Ass. U.K.* **32**: 497 - 520.

Parke, M. and Dixon, P.S. 1964. A revised checklist of British marine algae. *J. mar. biol. Ass. U.K.* **44**: 499 - 542.

Parke, M. and Dixon, P.S. 1968. Check-list of British marine algae - second revision. *J. mar. biol. Ass. U.K.* **48**: 783 - 832.

Parke, M. and Dixon, P.S. 1976. Check-list of British marine algae - third revision. *J. mar. biol. Ass. U.K.* **56**: 527 - 594.

Pilcher, B. 1967. The algae of John Templeton in the Ulster Museum. *Ir. Nat. J.* **15**: 350 - 353.

Praeger, R.Ll. 1938. *A Flora of the North-east of Ireland by S.A. Stewart and T.H. Corry.* Second edition. Belfast.

Praeger, R.Ll. 1949. *Some Irish Naturalists.* Dundalk.

South, G. and Tittley, I. 1986. *A Checklist and Distributional Index of the Benthic Marine Algae of the North Atlantic Ocean.* Huntsman Marine Laboratory and British Museum (Natural History), St Andrews and London.

Stewart, S.A. and Corry, T.H. 1888. *A Flora of the North-east of Ireland.* Belfast and Cambridge.

Wear, S. 1923. *A Second Supplement to and Summary of Stewart and Corry's Flora of the North-east of Ireland.* Belfast.

List of Plates

1. *"Fucus purpurascens"* Cystocloniun purpureum, cystocarpic. One of the fine illustrations in Templeton's *Hibernian Flora* MS. (Ulster Museum.)

2. *"Fucus purpurascens"* as noted in Templeton's *Hibernian Flora* MS. (Ulster Museum.)

3. *"Fucus clavellosa"* Lomentaria clavellosa. "On the shore of Glenarm July 14th 1808." Collected and determined by J. Templeton. (Ulster Museum: F22.)

4. William Thompson 1805-1852 (Ulster Museum.)

5. A typical sheltered boulder shore of Co. Down. Dominated by *Ascophyllum* and *Laminaria*. Downey's Rock, in Strangford Lough, 1990. (Photo. Julia D. Nunn, Ulster Museum.)

6. A moderately exposed rocky shore of Co. Antrim. Showing *Himanthalia elongata* and *Fucus serratus*. White Park Bay, Co. Antrim 1983. (Photo. Osborne Morton, Ulster Museum.)

7. Low littoral exposed shore showing *Laminaria* spp. Ballymacormick Point, Co. Down, 1993. (Photo. Julia D. Nunn, Ulster Museum.)

8. *Prasiola stipitata* at high water above the *Pelvetia* zone. Donaghadee, Co. Down 1984. (Photo. Osborne Morton, Ulster Museum.)

9. *Codium* sp. Darragh Island, Strangford Lough, Co. Down, 1987 (Photo. Julia D. Nunn, Ulster Museum.)

10. *Dictyopteris membranacea*. Sublittoral. East of Black Head, Rathlin Island, Co. Antrim. (Photo. B.E. Picton, Ulster Museum.)

11. Coralline rock pool dominated by *Corallina officinalis* with *Cladophora* sp. on an exposed shore. Portrush, Co. Antrim, 1989. (Photo. G.V. Day.)

12. *Lithothamnion glaciale* from a deep rock pool in the low littoral. Donaghadee, Co. Down, 1984. (Photo. Osborne Morton, Ulster Museum.)

13. *Stenogramme interrupta.* Sublittoral. Lee's Wreck, Strangford Lough, Co. Down, 1984. (Photo. Bernard E. Picton, Ulster Museum.)

14. *Schmitzia hiscockiana.* Loughan Bay near Torr Head,1983. Collected by Bernard E. Picton, determined by C.M. Howson. (Ulster Museum: F4368)

15. *Delesseria sanguinea.* Bloody Bridge, Co. Down, 1991. Collected and determined by Osborne Morton. (Ulster Museum: F8949)

16. *Heterosiphonia plumosa.* Sublittoral. Lee's Wreck, Strangford Lough, Co. Down. (Photo. Bernard E. Picton, Ulster Museum.)

Chlorophyta

Chlorophyceae

PRASIOLALES

Prasiola (C. Ag.) Menegh.

Prasiola furfuracea (Mert.) Kütz. **-A-**
Considered by some (*N A Checklist*) to be
synonymous with *P. stipitata*, EMB (*Sea
B I*) however gives it full species status.
Very rare, only one recent record.
ANTRIM - Carnlough 1983 on large
boulder at high water [OM det EMB]
F3650.

P. stipitata Suhr ex Jessen (Pl.8) **DA-**
Old records may include *P. calophylla*
(which may as a result be un-recorded).
Found in lichen zone above other algae on
rocky outcrops, often associated with
gull droppings.
DOWN and ANTRIM - Common.

Rosenvingiella Silva

Rosenvingiella polyrhiza (Rosenv.) Silva
DA-
The genus was origonally named *Gayella*
in 1893 and since then it has been
investigated a number of times. Edwards
(1975) indicated that *R. polyrhiza* may be
conspecific with *Prasiola stipitata*;
however it was listed as a seperate species
in *N A Checklist* and in *Sea B I*.
It is a small plant with a maximum length

of 14mm. Only four Irish records are
noted in Guiry (1978), none of which are
from NI.
All the records below are from the *NILS*.
DOWN - Swineley Pt 1985;
Ballymacormick Pt 1984; Orlock Pt 1985;
Whitechurch 1986; Robin's Rock 1986;
Green I 1986; Kearney 1985; Granagh B
1986; Kilclief Pt 1986; Ardglass B 1985;
Bloody Br 1985; Greencastle Rocks 1985.
ANTRIM - Rathlin I at Portandoon
1986 & Marie Isla 1986; Peggy's Hole
1985; Cushendun 1985; Red Arch 1985;
Garron Harbour 1985; Whitebay Pt 1986;
Black Hd 1986.

CHLOROCOCCALES

Characium A. Braun

Characium marinum Kjellm. **D—**
Not recorded in Britain since the turn of
the century and only recorded once in
Ireland. Epiphytic on various algae.
DOWN - Black Is 1983 [OM det EMB]
F3653c.

Chlorochytrium Cohn

[*C. inclusum* Kjellm. is now known to be a
phase in the life-history of
Spongomorpha aeruginosa qv].

Chlorochytrium sp.   **D—**
Unicellular, endophytic thallus. There
may be a number of unnamed species
(*Sea B I*). Apparently very rare, however
possibly not often looked for.
DOWN - Rolly I 1975 endophytic in
Enteromorpha flexuosa [OM det EMB]
F410; Black Is near Strangford
endophytic in *E. prolifera* 1983 [OM det
EMB] F3653.

C. cohnii Wright   **DA-**
This endophytic and microscopic alga is
included in *Br Check-list* but excluded
from the *N A Checklist*. Its life history is
incompletely known (*Sea B I*). Only
recorded from NI in the *NILS*.
DOWN - Yellow Rocks 1986; Holm B
1984; Ardglass B 1985; Rathmullan Pt
1985; Wreck Port 1985; S of Nicholson's
Pt 1985.
ANTRIM - Portandoon on Rathlin I 1986;
Gt Stookan 1986; Larry Bane B 1984;
Cushendun 1985; White Lady E 1985;
Marchburn Port 1984.

C. moorei Gardn.   **-A-**
(*C. willei*) Printz
Considered conspecific with *C. willei* and
a distinct species, not part of the life-
history of another green algae (*Sea B I*).
Very rare. Endophytic in *Blidingia
minima*. Only one record from NI.
ANTRIM - The Gobbins 1975 [OM det
EMB] F363b.

ULOTRICHALES

Ulothrix Kütz.

Ulothrix flacca (Dillw.) Thur.   **DAL**
(Incl *U. pseudoflacca* Wille and *U.
consociata* Wille)
Tufts of fine unbranched filaments; until
recently under-recorded as this species is
only recognisable under the microscope.
First recorded in NI in 1975 from Rolly I
and The Gobbins [OM conf EMB].

DOWN and ANTRIM - Common.
LONDONDERRY - Downhill Strand
1984 (*NILS*); Rinagree Pt 1984 (*NILS*).

U. implexa (Kütz.) Kütz.   **DAL**
(*U. subflaccida* Wille)
Small tufts of fine filaments. Previous to
NILS, only four records for all Ireland
from Cos Mayo, Cork and Dublin (MDG
1978).
All the records below are from the *NILS*.
DOWN - Ballymacormick Pt 1984;
Whitechurch 1986; Mahee I 1985;
Kircubbin Pt 1985; Yellow Rocks 1986;
Ringhaddy Rapids 1987; Chapel I
causeway 1986; Isle O'Valla 1986;
Ardglass B 1985; Rathmullan Pt 1985;
Bloody Br 1985; Wreck Port 1985;
Killough Harbour 1985.
ANTRIM - On Rathlin I at The Dutchman
1985; Port Gorm 1985; Port-na-Tober
1987; Portmore 1980's; Cushendun 1985;
Port Vinegar 1985; Red Arch 1985;
Turnly's Port 1985; Straidkilly Pt 1985;
Whitebay Pt 1986.

U. speciosa (Carm. ex Harv.) Kütz.   **DAL**
Rare in the literature, but recorded from
over 20 localities in Cos Down and
Antrim by *NILS*. Consisting of fine
unbranched filaments requiring
microscopic examination for
identification. Previous to the *NILS* it was
under-recorded.
First recorded in NI at Carnlough 1836
(*Ord S L*).
DOWN and ANTRIM - Common.
LONDONDERRY - New Br 1987
(*NILS*); Rinagree Pt 1984 (*NILS*).

Urospora Aresch.

Urospora bangioides (Harv.) Holm. et
Batt.   **-A-**
Apparently rare however possibly under-
recorded. It is regarded by some as a
growth form of *U. penicilliformis* and is
not included by EMB (*Sea B I*).

ANTRIM - Portballintrae pre-1871 (DM in *Phycol Brit*).

U. penicilliformis (Roth) Aresch.   **DAL**
Forming small tufts of simple filaments up to 8cm in length. Until recently much under-recorded.
DOWN - Culloden Hotel 1986; Barkley Rocks 1984; Whitechurch 1986; Wallaces Rocks 1984; Robin's Rocks 1986; Yellow Rocks 1986; Taggart I 1986; Granagh B 1986 (all *NILS*); Killough c.1934 (MJL & JMcG 1934); Rathmullan Pt 1985 (*NILS*); Bloody Br 1985 (*NILS*); Glasdrumman Port 1985 (*NILS*); Wreck Port 1985 (*NILS*); Greencastle Rocks 1985 (*NILS*).
ANTRIM - Common.
LONDONDERRY - New Br 1987 (*NILS*).

U. wormskioldii (Mert.) Rosenv   **-A-**
Simple tufted filaments up to 10cm long. Not recorded from Ireland previous to *NILS*. Rare.
All the records noted below are from *NILS*.
ANTRIM - Rathlin I at Killeany 1980's; Gt Stookan 1986; Port-na-Tober 1987; Carricknaford 1987; Dunineny Castle 1987; Ruebane Pt 1987; Portmore 1980's; Cushendun 1985; Campbeltown 1986; Whitebay Pt 1986.

Gomontia Born. et Flah.

Gomontia polyrhiza (Lagerh.) Born. et Flah.   **D/A-**
Very rare indeed, but possibly under-recorded. However it has been pointed out that at least six species of green shell-boring algae could be confused under this name.
DOWN/ANTRIM - Belfast L pre-1902 (*Batters' Cat*).
DOWN - "Co Down" 1898 (*Railway Guide*).

Monostroma Thur.

Monostroma grevillei (Thur.) Wittr. **DAL**
Similar to *Ulva lactuca,* but forming a delicate membrane one cell thick. Very rarely recorded except by *NILS,* overlooked by other recorders.
DOWN and ANTRIM - Common.
LONDONDERRY - Downhill Strand 1984 (*NILS*); Rinagree Pt 1984 (*NILS*).

M. obscurum (Kütz.) J. Ag.   **DA-**
(*Ulvaria obscura* (Kütz.) Gayral)
(*Monostroma fuscum* (Post. et Rupr.) Wittr.)
According to one reference (*Clare I S*) this species is abundant in Belfast L. However there are few records of it from NI.
DOWN/ANTRIM - Belfast L abundant pre-1915 (*Clare I S*).
DOWN - Ringhaddy Rapids 1987 (*NILS*); Wreck Port 1985 (*NILS*).
ANTRIM - Rathlin I at Doon B 1985 (*NILS*); The Burnfoot 1984 (*NILS*); Peggy's Hole 1985 (*NILS*); The Ladle 1985 (*NILS*); Whitebay Pt 1986 (*NILS*); Larne 1898 [HH].

M. oxyspermum (Kütz.) Doty   **DAL**
(*Ulvaria oxysperma* (Kütz.) Bliding)
Only recorded from NI by the *NILS*. There are two other records from elsewhere in Ireland.
DOWN - Culloden Hotel 1986; Swineley Pt 1986, Mid I B 1986; Ringhaddy Rapids 1987; Sheepland Harbour 1985; St John's Pt 1985; Craigalea 1980's.
ANTRIM - Rathlin I at: Stackamore 1987, Beddag 1985, Oweydoo 1987, The Dutchman 1985; The Burnfoot 1987; Peggy's Hole 1986; Carricknaford 1987; Larry Bane B 1984; Dunimeny Castle 1987; Ruebane Pt 1987; Cushendun 1985; Layd Church 1984; Campbeltown 1986; Whitebay Pt 1986; McIllroy's Port 1987; Chapman's Rock 1987.
LONDONDERRY - New Br 1987; Balls Pt 1987.

Spongomorpha Kütz.

Spongomorpha aeruginosa (L.) Hoek **DA-**
(*Chlorochytrium inclusum* Kjellm.)
  C. inclusum is now known to be a phase
in the heteromorphic life-history of *S.
aeruginosa* (*Sea B I*). It is a single celled
alga found within the thallus of *Dilsea
carnosa* and other algae. The other phase
is as small tufted filaments.
  DOWN - Carnalea 1916 [MWR]; near
Bangor 1925 (MC 1926);
Ballymacormick Pt 1984 (*NILS*); Orlock
Pt 1985 (*NILS*); Herring B 1985 (*NILS*);
Kircubbin Pt 1985 (*NILS*); Black Neb
(*NILS*); Horse I 1986 (*NILS*); Yellow
Rocks 1986 (*NILS*); Selk Rock 1986
(*NILS*); Chapel I causeway 1986 (*NILS*);
Kearney 1985 (*NILS*); Audley's Castle
rocks 1985 (*NILS*); Granagh B 1976 [OM
conf EMB as *C. inclusum*]; Kilclief Pt
1986 (*NILS*); Ardglass 1851 [WT] &
1985 (*NILS*); Bloody Br 1985 (*NILS*);
near St John's Pt 1988 [OM]; Newcastle
1838 [WT]; Glasdrumman Port 1985
(*NILS*); Wreck Port 1985 (*NILS*); S of
Nicholson's Pt 1985 (*NILS*).
  ANTRIM - Rathlin I pre-1851 (DM in
Phycol Brit), Marie Isla 1986 (*NILS*), E of
Lighthouse platform 1985 (*NILS*), Doon
B 1985 (*NILS*); Peggy's Hole 1985 &
1986 (*NILS*); Port Gorm 1985 (*NILS*);
The Ladle 1985 (*NILS*); Carricknaford
1987 (*NILS*); Red Arch 1985 (*NILS*).

S. arcta (Dillw.) Kütz.   **DAL**
  A common tufted alga 5-10cm in length.
First recorded in NI at Ballywalter 1836
by WT.
  DOWN and ANTRIM - Common.
LONDONDERRY - New Br 1987
(*NILS*); Rinagree Pt 1984 (*NILS*).

Capsosiphon Gobi

Capsosiphon fulvescens (C. Ag.) Setch. et
Gardn.   **DA-**
  Very rare, only recorded from NI in the

NILS.
DOWN - St John's Pt 1985.
ANTRIM - Straidkilly Pt 1985.

Blidingia Kylin

Blidingia marginata (J. Ag.) P. Dang.
                                               DAL
  Only recorded from NI by the *NILS*, the
only other Irish record being from Co
Wexford (TN 1970).
  DOWN - Foreland Pt 1986; Whitechurch
1986; Chapel I causeway 1986;
Carrstown Pt 1986; Killough Harbour
1985.
  ANTRIM - Portandoon on Rathlin I 1986;
Carrickmore 1985; Layd Church 1984;
Port Vinegar 1985; Garron Harbour 1985;
Turnly's Port 1985.
  LONDONDERRY - New Br 1987.

B. minima (Näg. ex Kütz.) Kylin   **DAL**
  A common but generally overlooked
species of the upper littoral.
  DOWN and ANTRIM - Common.
LONDONDERRY - New Br 1987 &
Downhill Strand 1984 both (*NILS*).

Enteromorpha Link

Enteromorpha clathrata (Roth) Grev.**DA-**
  Possibly a poorly recorded species
requiring further investigation.
  ANTRIM/DOWN - Belfast B (JT in
Harvey's Man).
  DOWN - Carnalea 1916 [MWR]; Bangor
pre-1811 (JT in *Eng Bot*), 1835 [WT];
Audley's Castle rocks 1985 (*NILS*);
Marlfield Rocks 1985 (*NILS*); Yellow
Rocks 1986 (*NILS*); Kircubbin Pt 1985
(*NILS*); Herring B 1985 (*NILS*); Long
Sheelah 1986 (*NILS*); Pawle I 1986
(*NILS*); Limestone Rock 1986 (*NILS*);
Portaferry dredged 1838 [WT]; Dundrum
1836 & 1851 [WT].

E. crinita (Roth) J. Ag.   **DAL**
(*E. ramulosa* (Sm.) Hook.)
  Uncommon, with few records mainly
  from Co Down.
  DOWN - Near Bangor 1835 (WT 1836);
  Ballymacormick Pt 1984 (*NILS*); Barkley
  Rocks 1984 (*NILS*); Herring B 1985
  (*NILS*); Long Sheelah 1986 (*NILS*); Pawle
  I 1986 (*NILS*); Holm B 1984 (*NILS*);
  Ballyhenry Pt 1975 [OM det EMB];
  Chapel I causeway 1986 (*NILS*); Ardglass
  B 1985 (*NILS*).
  ANTRIM - Port Vinegar 1985 (*NILS*);
  Red Arch 1985 (*NILS*).
  LONDONDERRY - Rinagree Pt 1984
  (*NILS*).

E. flexuosa (Wulf. ex Roth) J. Ag.   **DA-**
  Possibly conspecific with *E. lingulata* J.
  Ag. (*N A Checklist*) and is probably more
  common than indicated here.
  DOWN - Rolly I 1975 [OM det EMB];
  Ballyhenry B 1986 (*NILS*); Ballyquintin
  Pt 1972 [CRH det EMB]; Coney I 1985
  (*NILS*); N of Killowen 1985 (*NILS*).
  ANTRIM - Colliery B 1975 [OM det
  EMB]; Carrickfergus 1845 (WMcC in
  Phycol Brit).

E. intestinalis (L.) Link   **DAL**
  Considered by some as a good species
  separated from *E. compressa* by a
  sterility barrier (Larsen, 1981). However
  EMB (*Sea B I*) treats these as subspecies:-

  ssp. **intestinalis** (L.) Link - unbranched.
  (*E. intestinalis* (L.) Link)

  ssp. **compressa** (L.) Link - branched.
  (*E. compressa* (L.) Grev.)

  DOWN, ANTRIM and
  LONDONDERRY - Both subspecies are
  common.

E. linza (L.) J. Ag.   **DAL**
  A common and widespread species.
  First recorded in NI by JT at Connswater

pre-1825.
DOWN and ANTRIM - Common.
LONDONDERRY - Downhill Strand
1984 (*NILS*); Castlerock 1982 [OM];
Rinagree Pt 1984 (*NILS*).

E. prolifera (O.F. Müll.) J. Ag.   **DAL**
(*E. torta* (Mert.) Reinb.)
  EMB (*Sea B I*) considered *E. prolifera* to
  include *E. torta,* however other workers
  (Bliding, 1963) list them separately. *Sea B
  I* is followed here.
  DOWN and ANTRIM - Common.
  LONDONDERRY - New Br 1987
  (*NILS*); Downhill Strand 1984 (*NILS*);
  Rinagree Pt 1984 (*NILS*).

E. ralfsii Harv.   **D—**
  Very rare. Growing unattached in patches
  on mud in the littoral of sheltered shores.
  DOWN - Rough I in Strangford L 1975
  [OM]; Rolly I in Strangford L 1975 [OM
  det EMB as "probably"].

Percursaria Bory

Percursaria percursa (C. Ag.) Rosenv.
                                                DAL
  Filamentous threads of cells which are
  only distinctive under the microscope.
  Only three records prior to 1980 and few
  for all Ireland (MDG 1978). Probably
  under-recorded.
  DOWN - Swineley Pt 1986 (*NILS*);
  Ballymacormick Pt 1984 (*NILS*); Orlock
  Pt 1985 (*NILS*); Wallaces Rocks 1984
  (*NILS*); Marlfield Rocks 1985 (*NILS*);
  Selk Rock 1986 (*NILS*); Chapel I
  causeway 1986 (*NILS*); Kearney 1985
  (*NILS*); Isle O'Valla 1986 (*NILS*);
  Carrstown Pt 1986 (*NILS*); Ardglass B
  1985 (*NILS*); Newcastle 1838 [WT];
  Greencastle Rocks 1985 (*NILS*).
  ANTRIM - Portmore 1980's (*NILS*); Port
  Vinegar 1985 (*NILS*); Red Arch 1985
  (*NILS*); Ringfad 1984 (*NILS*);
  Drumnagreagh Port 1984 (*NILS*); Larne L
  1836 (*Ord S L*).

LONDONDERRY - Magilligan shore 1951 (HB 1951).

Ulva L.

Ulva lactuca L.   Sea Lettuce   **DAL**
A common species to be found on most rocky shores.
DOWN and ANTRIM - Common.
LONDONDERRY - Lower Doaghs 1985 (*NILS*); Downhill Strand 1984 (*NILS*); Castlerock 1982 [OM]; Rinagree Pt 1984 (*NILS*).

U. rigida C. Ag.   **DA-**
Very rare. Only one certain record.
DO N - Ardkeen 1975 [OM det EMB as: robably"].
A IM - Carnlough 1983 [OM det
F].

Bolbocoleon Pringsh.

leon piliferum Pringsh.   **D—**
 rare. Minute thalli creeping
 hytically or epiphytically on
 taria, and other algae EMB (*Sea B*

N - Sandeel B 1973 [OM det EMB]
07.

Entocladia Reinke

Entocladia flustrae (Reinke) W. Tayl. **DA-**
(*Epicladia flustrae* Reinke)
Small epizoic alga on the walls of polyzoa. Probably often overlooked.
DOWN - Common.
ANTRIM - Doon B on Rathlin I 1985 (*NILS*); Port Gorm 1985 (*NILS*); Port Vinegar 1985 (*NILS*); Drumnagreagh Port 1984 (*NILS*); McIllroy's Port 1987 (*NILS*); I of Muck 1987 (*NILS*); Dalaradia Pt 1986 (*NILS*); Black Hd 1986 (*NILS*); N side of Belfast L 1896 (TJ & RH 1896).

E. perforans (Huber) Levring   **DA-**
(*Epicladia perforans* Huber)

Minute filamentous alga forming inconspicuous patches in shells. No doubt under-recorded.
Only recorded from NI in *NILS*.
DOWN - Ballymacormick Pt 1984; Mahee I 1985; Kircubbin Pt 1985; Audley's Castle rocks 1985; Craiglewey 1984.
ANTRIM - Straidkilly Pt 1985; Marchburn Port 1984.

E. viridis Reinke   **DA-**
(*Phaeophila viridis* (Reinke) Burrows)
(*Acrochaete viridis* (Reinke) Nielsen)
Microscopic alga found on or in the cells of other algae. Probably not uncommon.
Only recorded in NI by the *NILS*.
DOWN - Culloden Hotel 1986; Swineley Pt 1986; Foreland Pt 1986; Barkley Rocks 1984; South I 1986; Mahee I 1985; Black Neb 1985; Robin's Rock 1986; Horse I 1986; The Dorn 1986; Taggart I 1986; Selk Rock 1980; Granagh B 1986; Mill Quarter B 1985; Rathmullan Pt 1985; Greencastle Rocks 1985.
ANTRIM - Rathlin I at:Maire Isla 1986, Stackamore & Oweydoo 1987; The Burnfoot 1984; Carricknaford 1987; Carrickmore 1985; Cave House shore 1984; Layd Church 1984; Port Vinegar 1985; Red Arch 1985; Old Pier 1985; Ringfad 1984; Campbeltown 1986; Straidkilly Pt 1985; Halfway House 1985; McIllroy's Port 1987; Riding Stone 1986; Black Hd E 1986.

E. wittrockii Wille   **DAL**
(*Phaeophila wittrockii* (Wille) Nielsen)
Microscopic alga in the cell walls of brown algae; similar to *E. viridis*. Only recorded in NI by the *NILS*.
DOWN - Ballymacormick Pt 1984; Orlock Pt 1985; Holm B 1984; Ballyquintin Pt 1986; Ardglass 1985; Coney I 1985; Rathmullan Pt & St John's Pt 1985.
ANTRIM - Rathlin I at : Marie Isla 1986, Stackamore 1987 & Doon B 1985;

Nº 67

Fucus purpurascens.

1. *"Fucus purpurascens" Cystoclonium purpureum,* cystocarpic. One of the fine illustrations in Templeton's *Hibernian Flora* MS. (Ulster Museum.)

Fucus Frond cylindrical

F. purpurascens. F. with a diaphanous purplish
brown irregularly branched frond
scattered setaceous ramuli, attenu
-ated at their base, acuminated
at their apices, and bearing soli
-tary, moniliform or confluent
spherical tubercles immersed in
their substance.

Fucus purpurascens. N. Ang
509 Bot Arr. 4.113 E. Bot. 1243 Lin
Tran. 3. 225 Syn. Fuc. 357. Hist Fuc 1.10.

F. tuberculatus H. Scot. 926
Gigartina purpurascens An. du Mus
20. 136.

Figured
E. Bot. t. 1243 imperfect. Hist Fuc. 1
t. 9 very perfect but less bush like
than it is often to be seen.

On Rocks in Pools on the rock
and often among the rejectamen
-ta.

This

2. *"Fucus purpurascens"* as noted in Templeton's *Hibernian Flora* MS. (Ulster Museum.)

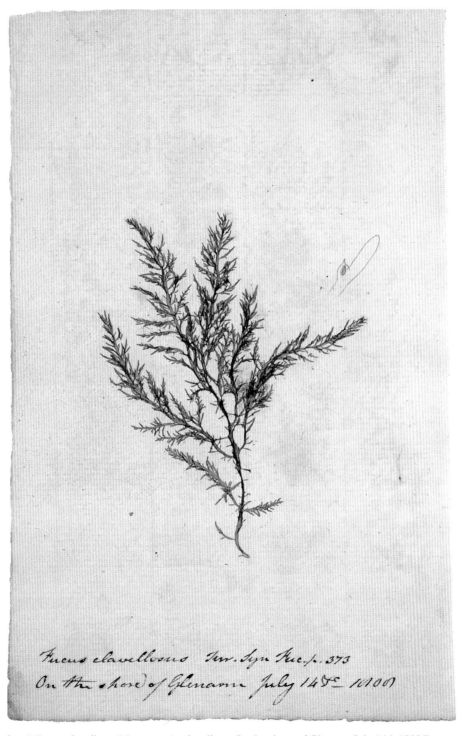

Fucus clavellosus Turn. Syn. Fuc.p. 373
On the shore of Glenarm July 14th 1808)

3. " *Fucus clavellosa* " *Lomentaria clavellosa*. On the shore of Glenarm July 14th 1808."
   Collected and determined by J. Templeton. (Ulster Museum: F22.)

4.   William Thompson 1805-1852. (Ulster Museum.)

5. A typical sheltered boulder shore of Co. Down. Dominated by *Ascophyllum* and *Laminaria*. Downey's Rock, in Strangford Lough, 1990. (Photo. Julia D. Nunn, Ulster Museum.)

6. A moderately exposed rocky shore of Co. Antrim showing *Himanthalia elongata* and *Fucus serratus*. White Park Bay, 1983. (Photo. Osborne Morton, Ulster Museum.)

7. Low littoral exposed shore showing *Laminaria* spp. Ballymacormick Point, Co. Down, 1993. (Photo. Julia D. Nunn, Ulster Museum.)

8. *Prasiola stipitata* at high water above the *Pelvetia* zone. Donaghadee, Co. Down, 1984. (Photo. Osborne Morton, Ulster Museum.)

Carrickarade 1986; Kinbane Hd 1985; Dunimeny Castle 1987; Ruebane Pt 1987; Cave House shore 1984; White Lady E 1985; Straidkilly Pt 1985; Riding Stone 1986; Black Hd E 1986.
LONDONDERRY - Rinagree Pt 1984.

Phaeophila Hauck

Phaeophila dendroides (P. et H. Crouan) Batt.   **-A-**
Very rare. Only three records in all Ireland, one of which is from NI the other two being from Cos Mayo and Wexford (MDG 1978). An epiphytic microscopic alga.
ANTRIM - Small sheltered bay at Cave House shore 1984 (*NILS*).

Pringsheimiella Hoehnel

Pringsheimiella scutata (Reinke) Marchew.   **DA-**
A very rare epiphyte growing as a disc up to 2mm in diameter on other alga.
DOWN - Swineley Pt 1986 (*NILS*); Wallaces Rocks 1984 (*NILS*); Granagh B 1986 (*NILS*); Ballyhenry B 1986 (*NILS*); The Dorn 1986 (*NILS*).
ANTRIM - Rathlin I at:Marie Isla 1986 (*NILS*), Beddag 1985 (*NILS*) & Oweydoo 1987 (*NILS*); Carricknaford 1987 (*NILS*); Fair Hd 1985 (*NILS*); Ringfad 1984 (*NILS*); Campbeltown 1986 (*NILS*); N side of Belfast L 1896 (TJ & RH 1896).

Ulvella P. et H. Crouan

Ulvella lens P. et H. Crouan   **DA-**
Very small epiphyte, less than 2mm diameter, growing on other algae, porcelain or glass. There are no Irish records previous to those below. All records are from the *NILS*.
DOWN - Ballymacormick Pt 1986; Orlock Pt 1985; Foreland Pt 1986; Barkley Rocks 1984; Wallaces Rocks 1984; South I 1986; Mahee I 1985;

Drummond I 1986; The Dorn 1986; Taggart I 1986; Ballyhenry B 1986; Kearney 1985; Isle O'Valla 1986; Ardglass B 1985; Bloody Br 1985; S of Nicholson's Pt 1985; Greencastle Rocks 1985.
ANTRIM - Rathlin I at : Marie Isla 1986 & Doon B 1985; The Burnfoot 1984; Port Gorm 1985; The Ladle 1985; Gt Stookan 1986; Port Moon B 1986; Carrickmore 1985; Fair Hd 1985; Portdoo 1986; Cave House 1984; White Lady 1985; Ringfad 1984; Carnfunnock B 1986; Black Hd 1986.

Tellamia Batt.

Tellamia contorta Batt.   **DA-**
(Incl *T. intricata* Batt.)
Minute filamentous alga growing on living marine snails but not penetrating the calcareous portion. Although EMB (*Sea B I*) considered *T. contorta* and *T. intricata* as separate species, South and Tittley (*N A Checklist*) is followed here. Until recently there were only eight records for both species for all Ireland, none of which were from NI (MDG 1978).
All the records below are from the *NILS*.
DOWN - Ballymacormick Pt 1984; Foreland Pt 1986; Barkley Rocks 1984; Wallaces Rocks 1984; South I 1986; Herring B 1985; Robin's Rock 1986; Ringburr Pt 1985; Selk Rock 1986; Granagh B 1986; Kilclief 1986; Mill Quarter B; Glasdrumman Port 1985; S of Nicholson's Pt 1985; Cranfield Pt 1985.
ANTRIM - Doon B on Rathlin I 1985; Peggy's Hole 1985; Carricknaford 1987; Portdoo 1986; Murlough B 1985; Old Pier 1984; Garron Harbour 1985; Turnly's Port 1985; Ringfad 1984; Campbeltown 1986; Straidkilly Pt 1985; Drumnagreagh Port 1984; Halfway House 1985; Seacourt 1985; Riding Stone 1986; Marchburn Port 1984; Loughshore Pk 1986.

CLADOPHORALES

Chaetomorpha Kütz.

Chaetomorpha crassa (C. Ag.) Kütz. **DA-**
(Poss conspecific with *C. linum*)
Not included in EMB (*Sea B I*) who
indicates that it may be synonymous with
C. linum. Only two old records of the
19th century.
DOWN - Dundrum B September 1851
[WT].
ANTRIM - Thompson's Bank, Belfast B
25 April 1836 [WT].

C. linum (O.F. Müll.) Kütz.   **DAL**
(*C. aerea* (Dillw.) Kütz.)
C. linum and *C. aerea* are considered as
different species by some authorities,
however EMB (*Sea B I*) regards them as
conspecific. *C. crassa* may also be
synonymous.
DOWN and ANTRIM - Common.
LONDONDERRY - Portstewart pre-1836
(DM in *Fl Hib*); Rinagree Pt 1984 (*NILS*).

C. mediterranea (Kütz.) Kütz.   **DA-**
(*C. capillaris* (Kütz.) Børg.)
Common, in places very common,
growing entangled on other algae in
littoral rock pools.
DOWN and ANTRIM - Common.

C. melagonium (Web. et Mohr) Kütz.**DAL**
Common, but never in abundance.
Usually only as scattered filaments.
Littoral and sublittoral to 12m.
DOWN, ANTRIM and
LONDONDERRY - Common.

Cladophora Kütz.

The recently confirmed records and those
records extracted from van den Hoek's
Revision...(*Rev Clado*) can be accepted with
confidence. However, as the species are
highly variable and the determination of
individual specimens is difficult, the older

records should be viewed with caution.

Cladophora albida (Huds.) Kütz.   **DAL**
Common in littoral rock pools.
DOWN - Common.
ANTRIM - Rathlin I at: Marie Isla 1986
(*NILS*), Stackamore 1987 (*NILS*),
Oweydoo 1987 (*NILS*) & Black Hd 1986
(*NILS*); Port Gorm 1985 (*NILS*); Giant's
Causeway pre-1841 (WT in *Harvey's
Man*); Larry Bane B 1984 (*NILS*);
Carrickarade 1986 (*NILS*) & Carrickmore
1985 (*NILS*); Murlough 1898 [**BEL**] &
Murlough B 1985 (*NILS*); Cushendun
1985 (*NILS*); Layd Church 1984 (*NILS*);
Red Arch (*NILS*); Garron Harbour 1985
(*NILS*); Ringfad 1984 (*NILS*); Seacourt
1985 (*NILS*); Brown's B - [C];
Chapman's Rock 1987 (*NILS*).
LONDONDERRY - Downhill Strand
1984 (*NILS*).

C. fracta (O.F. Müll. ex Vahl) Kütz.   **-A-**
Apparently very rare with few records,
none of which are recent.
ANTRIM - Antrim coast pre-1902
(*Batters' Cat*); Larne L 1937 (MJL
1960a).

C. hutchinsiae (Dillw.) Kütz.   **DAL**
Apparently rare in NI and with few recent
records.
ANTRIM/DOWN - First locks Lagan
1836 [WT]; Belfast B 1840 [GCH] & pre-
1841 (WT in *Harvey's Man*).
DOWN - Yellow Rocks 1986; Wreck Port
1985 both (*NILS*) records.
ANTRIM - Portrush 1836 (DM in *Fl
Hib*); Larne 1802 (JLD in *Harvey's Man*);
Belfast L pre-1853 (*Gifford's Mar Bot*).
LONDONDERRY - Magilligan 1871
(AM Irwin in HB 1951).

C. laetevirens (Dillw.) Kütz.   **D—**
Very rare indeed, only a few records the
older of which would benefit from
redetermination.
DOWN - Carnalea 1916 [MWR]; Black

Neb 1985 (*NILS*); Kilclief Pt 1986 (*NILS*); Newcastle 1839 [JJR?]; Dundrum B 1851 [WT].

C. lehmanniana (Liedenb.) Kütz.   **-A-**
Very rare, only one old record from NI.
ANTRIM - Murlough B 1898 [HH].

C. pellucida (Huds.) Kütz.   **DAL**
A not uncommon but readily identified species. Usually in shaded positions in large deep rock pools of the midlittoral to sublittoral depths of 8m.
ANTRIM/DOWN - Belfast L pre-1853 (*Gifford's Mar Bot*).
DOWN - Swineley Pt 1986 (*NILS*); Bangor 1835 [WT]; Luke's Pt 1982 [OM]; Ballymacormick Pt 1984 (*NILS*); Sandeel B 1973 [OM]; Orlock Pt 1985 (*NILS*); Lighthouse I 1975 & 1978 [OM]; Foreland Pt 1986 (*NILS*); near Donaghadee 1984 [OM]; Kinnegar 1983 [OM]; Barkley Rocks 1984 (*NILS*); near Ballywalter 1982 [*Sublit Surv*]; Ardkeen 1975 & 1976 [OM]; Ballyhenry I 1975 [OM]; Granagh B 1983 [*Sublit Surv*].
ANTRIM - Rathlin I at: E Lighthouse platform 1985 (*NILS*), Beddag 1985 (*NILS*), Doon B 1985 (*NILS*), The Dutchman 1985 (*NILS*), Killeany 1980's (*NILS*), Black Hd 1986 (*NILS*) & Arkill B 1984 [*Sublit Surv*]; Portrush 1915 [SW det OM]; Peggy's Hole 1985 (*NILS*); Port Gorm 1985 (*NILS*); near Ballintoy 1983 (*Sublit Surv*); Garron Pt 1835 [WT] & 1976 [OM]; Drumnagreagh Port 1983 [*Sublit Surv*]; Seacourt 1985 (*NILS*); Larne pre-1836 (JLD in WT 1836); Brown's B 1983 [*Sublit Surv*]; I of Muck 1987 (*NILS*).
LONDONDERRY - Rinagree Pt 1984 (*NILS*).

C. pygmaea Reinke   **DA-**
Very rare but possibly overlooked as it grows to less than 2mm high. Recorded in the sublittoral to depths of 25m.
DOWN - Strangford L narrows 1983 (*Sublit Surv*).
ANTRIM - Farganlack Pt & Cooraghy B on Rathlin I 1985 (*Sublit Surv*).

C. retroflexa (Bonnem. ex P. et H. Crouan) Hoek   **D—**
Only one record from NI.
DOWN - Craiglewey 1984 (*NILS*).

C. rupestris (L.) Kütz.   **DAL**
Very common, to be found in most rock pools of the littoral and into the sublittoral all year round.
First collected in NI at Portstewart 1834 (DM in *Rev Clado*).
DOWN, ANTRIM and LONDONDERRY - Common.

C. sericea (Huds.) Kütz.   **DAL**
Common, with a wide geographical range, from the Arctic to the Mediterranean and with a wide ecological range from sheltered shores to exposed shores. Capable of withstanding considerable fluctuations in salinity (*Rev Clado*).
DOWN and ANTRIM - Common.
LONDONDERRY - New Br 1987 (*NILS*); Downhill Strand 1984 (*NILS*).

C. vagabunda (L.) Hoek.   **-A-**
Very rare; only two records.
ANTRIM - Antrim coast pre-1902 (*Batters' Cat*); Greenisland 1857 (det EMH in *Rev Clado*).

Rhizoclonium Kütz.

Rhizoclonium tortuosum (Dillw.) Kütz.   **DAL**
Common. However until recently greatly under-recorded especially in Ireland. Filamentous.
First recorded in NI at the Giant's Causeway 1836 [WT].
DOWN and ANTRIM - Common.
LONDONDERRY - New Br 1987 (*NILS*); Downhill Strand 1984 (*NILS*);

Rinagree Pt 1984 (*NILS*).

CODIALES

Bryopsis Lamour.

Bryopsis hypnoides Lamour.   **DAL**
(Poss conspecific with *B. plumosa*.)
  Occasional, and rarer than *B. plumosa*, in
  shaded positions in low littoral rock
  pools and sublittoral to a depth of 8m.
  DOWN - Holywood 1840 [WT]; Helen's
  B 1989 [OM]; Holm B 1983 (*Sublit
  Surv*); Ballyhenry B 1984 [OM];
  Audley's Pt 1983 [*Sublit Surv*].
  ANTRIM - Rathlin I at Marie Isla 1986 &
  Doon B 1985 (*NILS*); Portrush pre-1836
  (DM in *Fl Hib*); Straidkilly Pt 1985
  (*NILS*); Halfway House 1985 (*NILS*);
  Larne 1841 [poss GCH det OM];
  Carrickfergus 1982 [OM]; Loughshore Pk
  1986 (*NILS*).
  LONDONDERRY - Derry prob pre
  c.1852 [Mr Moon]; S of Middle Bank
  1983 [*Sublit Surv*]; Castlerock 1982
  [OM].

B. plumosa (Huds.) C. Ag.   **DAL**
(Poss conspecific with *B. hypnoides*
Lamour.)
  Not uncommon in shaded positions in low
  littoral rock pools and sublittoral to 25m.
  Probably under-recorded.
  First specimen from NI was collected by
  JT at "The Whitehouse shore" in 1803.
  DOWN and ANTRIM - Common.
  LONDONDERRY - Moville Buoy L
  Foyle 1983 [*Sublit Surv*]; Downhill
  Strand 1984 (*NILS*); near Portstewart Pt
  1984 (*Sublit Surv*); near Rinagree Pt 1984
  (*Sublit Surv*) & (*NILS*).

Derbesia Solier

Derbesia marina (Lyngb.) Solier   **-A-**
(*Halicystis ovalis* (Lyngb.) Aresch.)
  H. ovalis is now known to be the haploid
  phase in the life-cycle of this species.

Very rare indeed, only one record.
ANTRIM - N side of Belfast L 1896, first
and only record for the NI (TJ & RH
1896).

Codium Stackh.

Codium adhaerens C. Ag.   **-A-**
  Extremely rare, only once recorded in NI
  and rarely in the rest of the British Is.
  ANTRIM - "... creeping over the
  limestone rocks in Church B in the island
  of Rathlin" 1837 (*Ord S L*). It shows an
  interesting record of existence on a small
  island in Western Invernesshire where it
  was first found in 1927 and refound in
  1965 on the same two adjacent boulders.
  However when OM visited Church B in
  1983 this species was not found, although
  other species of *Codium* were found.

C. bursa (L.) C. Ag.   **A\D-**
  Only recorded once in NI. As the
  whereabouts of that specimen is unknown
  and no other collector has found it in NI
  the record is considered doubtful.
  Nevertheless the species is not one easily
  confused with any other and recently
  (July 1977) two specimens were collected
  by diving in Co Donegal.
  ANTRIM/DOWN - "... near Belfast"
  between 1793 & 1810 (JT in *Eng Bot*).

C. fragile (Sur.) Hariot
  There are two subspecies of *C. fragile* and
  both occur in Ireland: ssp. *atlanticum* and
  ssp. *tomentosoides*. The latter is spreading
  rapidly and is very similar to ssp.
  atlanticum from which it is best
  distinguished microscopically by the tips
  of the utricles. In ssp. *atlanticum* the tip is
  blunt with no more than a slight
  protuberance, while in ssp. *tomentosoides*
  it is clearly pointed. The other similar
  species of *Codium*, *C. tomentosum* and *C.
  vermilara*, have rounded tips to the
  utricles.

ssp. **atlanticum** (Cotton) Silva   **DAL**
This is by far the most common
Codium in NI and although the other
subspecies is spreading it still is less
common.
DOWN - Carnalea 1946 [MPHK det
HMP]; Luke's Pt 1982 drift (OM);
Ballymacormick Pt 1984 (*NILS*),
1988 [OM]; Orlock Pt 1985 (*NILS*);
Sandeel B 1972 & 1973 [OM conf
HMP]; Foreland Pt 1986 (*NILS*); near
Donaghadee 1967 [OM conf HMP];
Barkley Rocks 1984 (*NILS*);
Whitechurch 1986 (*NILS*); Wallaces
Rocks 1984 (*NILS*); Mahee I 1985
(*NILS*); Horse I 1986 (*NILS*); Castle I
Pt 1986 (*NILS*); Portavogie 1988
(OM); Selk Rock 1986 (*NILS*);
Ballyhenry I 1975 [OM]; Kearney
1985 (*NILS*) & 1987 [OM]; Chapel I
causeway 1986 (*NILS*); Tara Pt 1953
(CID det PCS in PCS 1955);
Carrstown Pt 1986 (*NILS*); Kilclief Pt
1986 (*NILS*); Ballyquintin Pt 1986
(*NILS*); Mill Quarter B 1985 (*NILS*);
Bloody Br 1985 (*NILS*); Ballyhornan
1982 [OM]; Rathmullan Pt 1985
(*NILS*).
ANTRIM - Rathlin I at Marie Isla
1986 (*NILS*) & The Dutchman 1985
(*NILS*); Doon B 1985 (*NILS*); Church
B 1983 [OM]; Portrush 1858
(Hennedy det PCS 1955), 1915 [SW
conf HMP]; Port Gorm 1985 (*NILS*);
Portballintrae 1960 [MPHK det
HMP]; Giant's Causeway 1910 (ADC
det PCS 1955); Port Moon B 1986
(*NILS*); Colliery B 1978 [OM];
Carrickmore 1985 (*NILS*); Cave
House shore 1984 (*NILS*); Layd
Church 1984 (*NILS*); Port Vinegar
1985 (*NILS*); Garron Harbour 1985
(*NILS*); Ringfad 1984 (*NILS*);
Ballygalley Hd 1986 (*NILS*); Larne
1865 (CA Johnson det PCS 1955);
Riding Stone 1986 (*NILS*); Dalaradia
1986 (*NILS*); The Gobbins 1975 [OM
conf HMP]; Carrickfergus 1840 (J

Doran det PCS 1955).
LONDONDERRY - Downhill Strand
1984 (NILS); Portstewart (DM);
Rinagree Pt 1984 (*NILS*).

ssp. **tomentosoides** (van Goor) Silva
D—
DOWN - Herring B 1985 (*NILS*);
Rolly I 1975 drift [OM]; Kircubbin
1985 (*NILS*), 1988 & 1990 [OM];
Black Neb inlet 1986 (*NILS*); Mid I
B 1986 (*NILS*); Yellow Rocks 1986
(*NILS*); Ardkeen 1975 [OM];
Granagh B 1972 [RKB det HMP];
Carrstown Pt 1986 (*NILS*); Bar Hall
B 1986 (*NILS*); Bloody Br 1990 [PH
det OM].
ANTRIM and LONDONDERRY -
No records as yet.

C. tomentosum Stackh.   **-AL**
Rather rare. The old records, unless
supported by confirmed specimens,
cannot be accepted with confidence.
ANTRIM - Rathlin I at Doon B 1985
(*NILS*) & Black Hd 1986 (*NILS*); Portrush
1903 (JA 1904b); White Rocks 1979
[OM]; Bushmills 1833 [GCH det HMP];
Ballycastle 1797-98 (RB det PCS 1955),
1903 (JA 1904b); Colliery B 1898 [HH
conf HMP], 1978 [OM]; Cushendall 1903
(JA 1904b); Blackcave Tunnel 1903 (JA
1904b); Larne pre-1836 (JLD 1837).
LONDONDERRY - Black Rock 1915
[SW det HMP], 1981 [OM]; Rinagree Pt
1984 (*NILS*).

C. vermilara (Olivi) Chiaje   **-A-**
Rare, however posssibly confused with
other species of *Codium*.
ANTRIM - Portrush 1835 (DM) &
Cushendall pre-1866 (WHH both in PCS
1955); Portmuck 1974 [PH det HMP].

Blastophysa Reinke

Blastophysa rhizopus Reinke   **-A-**
Extremely rare; endophytic in larger

algae. Only recorded once in NI and rare
in the British Is. Minute and possibly
overlooked.
ANTRIM - Torr Hd 1898 [HH].

Phaeophyta

Phaeophyceae

ECTOCARPALES

Acinetospora Born.

Acinetospora crinita (Carm. ex Harv.) Kornm. **DA-**
A small tufted filamentous alga unknown from NI until recently. Rare. All the records below are from the *NILS*.
DOWN - Orlock Pt 1985; Castle I Pt 1985; Taggart I 1986; Selk Rock 1986; Chapel I causeway 1986; Kearney 1985; Green I 1986; Bar Hall B 1986; Audley's Castle rocks 1985; Greencastle Rocks 1985.
ANTRIM - Black Hd E 1986.

Ectocarpus Lyngb.

Ectocarpus fasciculatus Harv. **DAL**
A widespread and common species, until recently considerably under-recorded. Small and similar to *E. siliculosus* and other small algae, requiring microscopic examination for identification. Usually epiphytic on the larger fucoids and laminarians.
DOWN and ANTRIM - Common.
LONDONDERRY - Rinagree B 1983 (OM).

E. siliculosus (Dillw.) Lyngb. **DAL**
A widespread and common species

resembling *E. fasciculatus* and until recently under-recorded. Commonly epiphytic on other algae.
DOWN and ANTRIM - Common.
LONDONDERRY - Rinagree Pt 1984 (*NILS*).

Feldmannia Hamel

Feldmannia globifera (Kütz.) Hamel **DA-**
A rare small tufted alga only recently recorded from NI by the *NILS*.
DOWN - Granagh B 1986 (*NILS*).
ANTRIM - Ringfad 1984 (*NILS*).

Giffordia Batt.

Giffordia granulosa (Sm.) Hamel **DA-**
Probably not uncommon but like other small filamentous algae it is likely to be under-recorded.
DOWN - Swineley Pt 1986 (*NILS*); Luke's Pt 1982 [OM]; Sandeel B 1972 (OM conf GR in OM 1974); Barkley Rocks 1984 (*NILS*); near Rough I 1983 [OM]; Ballywalter 1984 [OM]; Long Sheelah 1986 (*NILS*); Ardkeen 1975 [OM conf GR]; Green I 1986 (*NILS*); Ballyquintin Pt 1986 (*NILS*); Sheepland Harbour 1985 (*NILS*); Dundrum Inner B 1987 (*NILS*).
ANTRIM - Peggy's Hole 1985 (*NILS*); Larry Bane Hd 1985 (*NILS*); Ballycastle

pre-1902 (*Batters' Cat*); Riding Stone 1986 (*NILS*); I of Muck E 1987 (*NILS*); Barney's Pt 1986 (*NILS*); Carrickfergus 1982 [OM]; Belfast L pre-1902 (*Batters' Cat*).

G. hincksiae (Harv.) Hamel   **DAL**
Until recently *G. hincksiae* was considered very rare in Ireland.
First recorded as an epiphyte on *Saccorhiza polyschides* at Ballycastle where it was found in 1840 by Miss Hincks [**BEL**] after whom it was named (*Phycol Brit*).
All the records below are from the *NILS*.
DOWN - Ballymacormick Pt 1984; Foreland Pt 1986; Down Rock 1986; Carrstown Pt 1986; Ballyquintin Pt 1986; Bloody Br 1985 & S of Nicholson's Pt 1985.
ANTRIM - Common.
LONDONDERRY - Rinagree Pt 1984.

G. ovata (Kjellm.) Kylin   **D—**
Very rare. Small filamentous alga, only recorded in Ireland by the *NILS*.
DOWN - Ballymacormick Pt 1984; Selk Rock 1986; Chapel I causeway 1986.

G. sandriana (Zanard.) Hamel   **D—**
Very rare. This is the first record for Ireland to my knowledge.
DOWN - Drummond I 1986 (*NILS*).

G. secunda (Kütz.) Batt.   **DA-**
Very rare; only three records from NI and all from the *NILS*.
DOWN - The Dorn 1986.
ANTRIM - Rathlin I at Marie Isla 1986; Layd Church 1984.

Herponema J. Ag.

Herponema velutinum (Grev.) J. Ag. **D—**
Herponema is a very rare genus in Ireland with no published records from NI.
However there is one specimen in **BEL** with the name: "*Sph velutina*", one of the

old synonyms for the species.
DOWN - Strangford L 1835 [WT] requires confirmation.

Laminariocolax Kylin

Laminariocolax tomentosoides (Farlow) Kylin   **DAL**
A small filamentous alga unknown in NI previous to the *NILS*.
All the records below are from the *NILS*.
DOWN - Swineley Pt 1986; Ballymacormick Pt 1984; Herring B 1985; Mahee I 1985; Ballyquintin Pt 1986; Sheepland Harbour 1985; Craigalea 1980's; St John's Pt 1985; Glasdrumman Port 1985.
ANTRIM - Rathlin I at: Beddag 1985, The Dutchman 1985 & Killeany 1980's; The Burnfoot 1984; Peggy's Hole 1986; Port Moon B 1986; Carricknaford 1987; Carrickarade 1986; Fair Hd W 1985; Murlough B 1985; Portmore 1980's; Cushendun 1985; Garron Harbour 1985; Turnly's Port 1985; Carnfunnock B 1986; Chapman's Rock 1987.
LONDONDERRY - Rinagree Pt 1984.

Phaeostroma Kuck.

Phaeostroma pustulosum Kuck.   **-A-**
Very rare, only one record from NI and few from all of Ireland.
ANTRIM - Murlough 1898 [HH].

Pilayella Bory

Pilayella littoralis (L.) Kjellm.   **DAL**
A very common filamentous littoral alga, usually to be found growing epiphytically on *Fucus* or *Ascophyllum*.
DOWN and ANTRIM - Common.
LONDONDERRY - New Br 1987 (*NILS*); Balls Pt 1987 (*NILS*); Lower Doaghs 1984 (*NILS*); Downhill Strand 1984 (*NILS*); Rinagree Pt 1984 (*NILS*).

Spongonema Kütz.

Spongonema tomentosum (Huds.) Kütz.
DAL
A common epiphyte of *Fucus*,
Himanthalia and other large algae in the
littoral.
First recorded in NI from Larne in 1827
[JLD].
DOWN and ANTRIM - Common.
LONDONDERRY - Rinagree Pt 1984
(*NILS*).

Pseudolithoderma Svedelius

Pseudolithoderma extensum (P. et H.
Crouan) Lund **DA-**
Apparently rare, but probably under-
recorded. Encrusting on rock in the low
littoral into the sublittoral to a depth of
25m.
DOWN - South I 1986 (*NILS*); Wallaces
Rocks 1984 (*NILS*); Black Neb 1985
(*NILS*); W of Rue Pt, Rue Pt & Cloghy
Rocks in Strangford L narrows 1983
(*Sublit Surv*).
ANTRIM - Rathlin I at Farganlack Pt &
Cooraghy B 1983 (*Sublit Surv*).

Ralfsia Berk.

Ralfsia verrucosa (Aresch.) J. Ag. **DA-**
A widespread and common encrusting
alga. However it is inconspicuous and
similar to *Stragularia*.
Until this century only recorded once in
NI at Annalong 1851 [WT].
DOWN and ANTRIM - Common.

Stragularia Strömfelt

Stragularia clavata (Harv.) Hamel **DAL**
(*Ralfsia clavata* (Harv.) P. et H. Crouan)
Laboratory cultures suggest that this
crustose alga may be involved in the life-
history of *Petalonia*. However this
requires further investigation as it appears
to be determined by environmental

conditions (*Sea B I*). Widespread and
common, but dies back during the
summer and is similar to *Ralfsia*.
All the records below are from the *NILS*
save the one detailed.
DOWN - Ballymacormick Pt 1984;
Barkley Rocks 1984; Wallaces Rocks
1984; South I 1986; Mahee I 1985;
Herring B 1985; Black Neb 1984; Castle I
Pt 1986; Ballyhenry B 1986; Strangford
Harbour 1985 [CAM det RLF];
Ballyquintin Pt 1986; Mill Quarter B
1985; Craiglewey 1984; Wreck Port
1985; Greencastle Rocks 1985; Cranfield
Pt 1985.
ANTRIM - Rathlin I at : Portandoon
1986, Marie Isla 1986 & Portawillin
1986; The Burnfoot 1984; Port Gorm
1985; Port Moon B 1986; Carrickarade
1986; Fair Hd W 1985; Layd Church
1984; Red Arch 1985; White Lady 1985;
Ringfad 1984; Campbeltown 1986;
Drumnagreagh Port 1984; Black Hd
1986; Loughshore Pk 1986.
LONDONDERRY - Rinagree Pt 1984.

Myrionema Grev.

Myrionema strangulans Grev. **DA-**
A common epiphyte on *Ulva* and
Enteromorpha, however the records of it
are few as the species is small and easily
overlooked.
DOWN - Sandeel B 1972 (OM) & 1973
[OM]; near Ballywalter 1985 [OM]; Rolly
I 1975 [OM conf GR]; Kircubbin Pt 1985
(*NILS*); Yellow Rocks 1986 (*NILS*);
Ardkeen 1976 [OM]; Portaferry 1837
[WT]; Mill Quarter B 1985 (*NILS*).
ANTRIM - Giant's Causeway 1839
[WT]; White Pk B 1915 [MWR];
Murlough B 1898 [HH].

Ulonema Fosl.

Ulonema rhizophorum Fosl. **DA-**
Some workers consider that this should be
included within *Myrionema strangulans*.

It is, however, retained here as a separate species pending further investigation. Said to be common and appearing as spots growing epiphytically on *Dumontia contorta*. However records of it are few and require confirmation. Possibly overlooked and under-recorded due in part to its small size.
DOWN - Sandeel B 1973 [OM] F3076.
ANTRIM - Portbraddan 1983 [OM] F3706.

Elachista Duby

Elachista flaccida (Dillw.) Aresch.   **D—**
Small epiphytic tufts on *Cystoseira* sp. in low littoral pools and sublittoral. Only recorded once in NI.
DOWN - Chapel I causeway 1986 (*NILS*).

E. fucicola (Vell.) Aresch.   **DAL**
A small common epiphyte on *Fucus*.
DOWN and ANTRIM - Common.
LONDONDERRY - New Br 1987 (*NILS*); Balls Pt 1987 (*NILS*); "... dredged Magilligan, Culkeeragh" c.1937 - 48 (HB); Castlerock 1982 (OM); Rinagree Pt 1984 (*NILS*).

E. scutulata (Sm.) Duby   **DA-**
Epiphytic on *Himanthalia* as small tufts in the low littoral and upper sublittoral of the more exposed localities.
All the records are from the *NILS*.
DOWN - Kearney 1985; Granagh 1986; Kilclief 1986.
ANTRIM - Rathlin I at: Portandoon 1986, Black Hd 1986, Killeany 1980's & Marie Isla 1986; Port Gorm 1985; Carricknaford 1987; Kinbane Hd 1985; Cave House shore 1984; Ringfad 1984; Straidkilly Pt 1985.

Leptonematella Silva

Leptonematella fasciculata (Reinke) Silva
                                              D—
A small epiphytic or epizoic alga, forming

tufts of filaments to 12mm long. Only one previous record from Co Cork in Ireland.
DOWN - Kearney 1985 (*NILS*).

CHORDARIALES

Petrospongium Näg.

Petrospongium berkeleyi (Grev.) Näg. **-A-**
(*Cylindrocarpus berkeleyi* (Grev.) P. et H. Crouan)
Very rarely recorded in NI and all records old. However as it forms only small (10 - 20mm) cushions it is probably overlooked and under-recorded. Harvey (*Phycol Brit*) records it as plentiful on the W coast and probably "pretty generally distributed along our shores, being overlooked on account of its being nearly of the colour of the rock ... and resembling, in its fleshy appearance and feel, the collapsed body of the common *Actinia*" (sea anemone).
ANTRIM - Murlough B 1898 [HH], this was published as "Murlough B a little to the east of Miss Charle's cottage" 1899 (HH); Giant's Causeway 1899 (HH).

Leathesia S.F. Gray

Leathesia difformis (L.) Aresch.   **DAL**
A common epiphytic species of the littoral.
First recorded in NI at Ballywalter by WT in 1836.
DOWN and ANTRIM - Common.
LONDONDERRY - Rinagree Pt 1984 (*NILS*).

Myriactula Kuntze

Myriactula areschougii (P. et H. Crouan) Hamel   **-A-**
Only one old record, very rare. A minute epiphyte in the conceptacles of *Himanthalia elongata*.
ANTRIM - First record in Ireland at Murlough B 1898 (HH in Anon 1899b).

Spermatochnus Kütz.

Spermatochnus paradoxus (Roth) Kütz.
D—
Very rare, only two records. Sublittoral.
DOWN - Strangford L pre-1902 (*Batters' Cat*); Marlfield B 1983 (*Sublit Surv*).

Stilophora J. Ag.

Stilophora rhizodes (Turn.) J. Ag. **DA-**
Rare in NI with no recent records. A southern species (*Prov Atlas*).
ANTRIM/DOWN - Belfast L pre-1841 (WT in *Harvey's Man*).
DOWN - Ballyholme B "among rejectamenta" 1812 [WT]; Bangor "washed ashore in considerable plenty" 1835 (WT 1836); there is also an unidentified herbarium specimen [BEL] from: "Bangor 1835 WT" which is almost certainly *S. rhizodes*; it is also recorded as dredged in Strangford L (4 fathoms) 1851 [WT]; Portaferry 1837 [WT].
ANTRIM - Nearest coast ... to The Maiden Rocks pre-1871 (GD 1871); Belfast L N side 1896 (TJ & RH 1896).

Acrothrix Kylin

Acrothrix gracilis Kylin **D—**
Very rare, only two records from NI both from the *NILS*.
DOWN - Barkley Rocks 1984 & Wallaces Rocks 1984.

Chordaria C. Ag.

Chordaria flagelliformis (O.F. Müll.) C. Ag. **DA-**
A fairly common species in the littoral of most rocky shores.
ANTRIM/DOWN - Near Belfast pre-1825 (JT in *Alg Brit*).
DOWN - Holywood 1836 (JSD); Bangor 1915 [SW]; Donaghadee 1910 (JA 1913) & 1967 [OM]; Strangford L 1851 [WT]; Rolly I 1977 [OM]; Black Neb 1986

(*NILS*); Portavogie 1915 [MWR]; Yellow Rocks 1986 (*NILS*); The Dorn 1986 (*NILS*); Ringburr Pt 1985 (*NILS*); Coney I 1985 (*NILS*); Newcastle 1808 [JT]; Bloody Br 1985 (*NILS*); Wreck Port 1985 (*NILS*); Greencastle 1856 [WS].
ANTRIM - Portandoon on Rathlin I 1986 (*NILS*); Portballintrae 1960 [MPHK]; Ballycastle pre-1871 (WH in *UCC Cat*); Colliery B 1978 [OM]; Murlough B 1898 [HH]; Cushendall pre-1866 [WHH]; Ringfad 1978 [OM]; Cairnlough 1836 (JSD 1837); Larne L 1802 [JLD det JT] & pre-1852 [JLD]; Carrickfergus pre-1852 [BEL] & 1931 [RN Gregg]; Macedon Pt 1915 [SW].

Eudesme J. Ag.

Eudesme virescens (Carm. ex Harv.) J. Ag. **DA-**
Rather rare judging from the few recent records.
DOWN - Near Craigavad pre-1825 [JT det MWR]; Carnlea 1916 [MWR]; Bangor 1831 [JLD]; Yellow Rocks 1986 (*NILS*); Wreck Port 1985 (*NILS*).
ANTRIM - Ballycastle pre-1801 [RB det MWR]; Colliery B 1898 [HH]; Garron Pt 1976 [OM]; Glenarm 1808 [JT det MWR]; Larne 1808 [JT det MWR] & pre-1836 (JLD).

Mesogloia C. Ag.

Mesogloia vermiculata (Sm.) S.F. Gray **DA-**
This species is said to be common on the NE coast of Ireland (*Batters' Cat*) and "Mr Thompson" is said to have found it in profusion pre-1857 (*Syn Seaw*). However there are no recent records of it at all and it would seem now to be rare.
DOWN - Bangor 1804 [JT]; Newcastle 1808 [JT]; Annalong 1851 [WT].
ANTRIM - Cushendall pre-1866 [WHH]; shores about Glenarm, Larne etc 1808 [JT], near Larne 1806 (JLD in *Eng Bot*);

nearest coast to The Maiden Rocks pre-1871 (GD).

Sphaerotrichia Kylin

Sphaerotrichia divaricata (C. Ag.) Kylin
-A-
Very rare, only recorded once. This single record has been quoted in most of the standard references and appears to be the only record of the species in all Ireland.
ANTRIM - All the published records appear to be based on one find by W. McCalla, for example: "thrown up from deep water, at Carrickfergus, near Belfast, Mr. McCalla Oct 1845" (*Phycol Brit*), 1845 [WMcC].

CUTLERIALES

Cutleria Grev.

Cutleria multifida (Sm.) Grev.   **DA-**
(Incl *Aglaozonia parvula* (Grev.) Zanard.)
The full life history of this species is heteromorphic. However it seems that it is a "flexible" one, for the *Aglaozonia* or prostrate phase, is much more common in NI than the erect bladed *Cutleria* phase.

Sexual (*Cutleria*) phase.
Almost unknown in the littoral and very rare in the sublittoral.
DOWN - Ballymacormick Pt 1988 [OM]; near Ballywaddan 1988 [OM]; SW of Dunnyneill Is 1983 [*Sublit Surv*]; Audley's Pt 1983 [*Sublit Surv*]; S of Cow & Calf 1984 [*Sublit Surv*]; Annalong 1982 [*Sublit Surv*].

Asexual (*Aglaozonia*) phase.
DOWN and ANTRIM - Common in the sublittoral to 25m especially in the more explored sites such as Rathlin I (*Sublit Surv*).

TILOPTERIDALES

Tilopteris Kütz.

Tilopteris mertensii (Turn.) Kütz.   **DA-**
A very rare species of the low littoral and sublittoral.
DOWN - Strangford L dredged pre-1841 (WT in *Harvey's Man*); near Portaferry dredged 1838 [WT].
ANTRIM - Carrickfergus pre-1846 (WMcC in *Phycol Brit*).

SPHACELARIALES

Sphacelaria Lyngb.

Sphacelaria bipinnata (Kütz.) Sauv.   **D—**
One record only from NI.
DOWN - Herring B 1985 (*NILS*).

S. cirrosa (Roth) C. Ag.   **DAL**
Generally considered to be common, even abundant, however the records are few for Cos Antrim and Londonderry. Probably under-recorded. Low littoral and sublittoral.
DOWN - Common.
ANTRIM - Rathlin I at Black Hd 1986 (*NILS*).
LONDONDERRY - Castlerock 1982 [OM]; Black Rock 1981 [OM].

S. fusca (Huds.) S.F. Gray   **DAL**
Records previous to 1980's are few and suggested that this was a rare species. However the recent *NILS* shows it to be more common and probably previously overlooked.
DOWN and ANTRIM - Common.
LONDONDERRY - Rinagree Pt 1984 (*NILS*).

S. plumosa Lyngb.   **D[A]-**
Rare but inconspicuous and possibly overlooked. Sublittoral to depths of at least 4m.
ANTRIM/DOWN - Belfast B 1840 [EG],

pre-1841 (WT in *Harvey's Man*).
DOWN - Ballymacormick Pt 1984
(*NILS*); Donaghadee dredged 1842
[GCH]; Ballywalter 1836 [WT];
Drummond I 1986 (*NILS*); Portaferry pre-
1902 (*Batters' Cat*); Rathmullan Pt 1985
(*NILS*); Craigalea 1980's (*NILS*); Bloody
Br 1985 (*NILS*); Glasdrumman Port 1985
(*NILS*); Mullartown Pt 1982 [*Sublit Surv*];
Wreck Port 1985 (*NILS*); Murphy's Pt
1982 [*Sublit Surv*]; Carlingford L 1984
[*Sublit Surv*].

S. radicans (Dillw.) C. Ag.   **DAL**
Until *NILS* this species was unrecorded
from NI (MDG 1978). However it
appears to have been overlooked,
unrecorded and possibly confused with
other species of the genus.
All the records below are from the *NILS*.
DOWN - Ballymacormick Pt 1984;
Orlock Pt 1985; Barkley Rocks 1984;
Wallaces Rocks 1984; Mid I B 1986; Selk
Rock 1986; Chapel I causeway 1986;
Kearney 1985; Rathmullan Pt 1985;
Craigalea 1980's; Bloody Br 1985; Wreck
Port 1985; N of Killowen 1985.
ANTRIM - Rathlin I at Portandoon 1986
& E of Lighthouse platform 1985; Cave
House shore 1984; Red Arch 1985;
Ringfad 1985; Black Hd E 1986.
LONDONDERRY - Rinagree Pt 1984;
Downhill Strand 1984.

Halopteris Kütz.

Halopteris filicina (Grat.) Kütz.   **DAL**
Rare in the literature but not uncommon
especially in the sublittoral down to 28m.
Most of the records from Co Antrim.
ANTRIM/DOWN - Belfast B pre-1841
(WT in *Harvey's Man*).
DOWN - Bangor 1804 [JT] & Ballyholme
1835 [GCH & WT]; near Copeland Is
1985 (*Sublit Surv*); near Ballywalter 1982
[*Sublit Surv*]; near Colin Rock 1983
[*Sublit Surv*]; Lee's Wreck 1984 [*Sublit
Surv*]; Granagh B 1983 [*Sublit Surv*].

ANTRIM - Portrush 1856 [WS det OM];
Rathlin I at 18 sites 1983, 1984, 1985
[*Sublit Surv*]; Ballycastle at 3 sites 1985
(*Sublit Surv*); Carrickarade I 1985 (*Sublit
Surv*); Torr Hd 1983 (*Sublit Surv*); Red B
1983 (*Sublit Surv*); Ringfad Pt 1983
(*Sublit Surv*); Drumnagreagh Port 1983
[*Sublit Surv*]; 3 sites at Ballygalley Hd
1983, 1985 (*Sublit Surv*); Whitehouse Pt
1807 [JT det OM].
LONDONDERRY - Black Rock 1915
[SW det OM].

H. scoparia (L.) Sauv.   **D—**
A very rare southern species. No
published records from NI. Low littoral
and sublittoral.
DOWN - Only one doubtful record:
Bangor 1835 [WT].

Cladostephus C. Ag.

Cladostephus spongiosus (Huds.) C. Ag.
A common species in pools of the littoral
and upper sublittoral. This species exists
in two forms which, until recently, were
considered separate species.

f. spongiosus   **DAL**
(*C. spongiosus* (Huds.) C. Ag.)
   DOWN - Holywood 19th C [anon];
   Craigavad 1837 [WT]; Rockport
   1979 (OM); near Wilson's Pt 1930
   (MC 1930); Bangor 1837 [WT], prob
   drift 1915 [SW]; Sandeel B 1972
   (OM 1974); Rolly I 1975 (OM);
   Ardkeen 1976 (OM); St John's Pt
   1984 [*Sublit Surv*]; Newcastle 1836
   [WT]; Annalong 1851 [WT]; Kilkeel
   1985 (*Sublit Surv*); Vidal Rock 1984
   [*Sublit Surv*].
   ANTRIM - North B Portrush 1915
   [SW], Portrush 1930 (WSW & JWW
   1933a), 1949 [WRM det HB]; White
   Rocks 1979 (OM) & 1980 [OM];
   Dunluce c.1930 (WSW & JWW
   1933b); Portballintrae 1960
   [MPHK]; Colliery B 1978 [OM];

Garron Pt 1976 (OM); Carnlough 1983 (OM); Glenarm B 1983 (OM); Ballygalley 1975 (OM); Portmuck 1975 [OM]; The Gobbins 1898 [HH]; Carrickfergus 1982 (OM). LONDONDERRY - Castlerock 1982 (OM); Black Rock 1915 [SW].

f. verticillatus (Lightf.) P.v. R.   **DAL**
(*C. verticillatus* (Lightf.) C. Ag.)
Apparently a relatively rare form compared with f. *spongiosus* with which it may be confused. There are few recent records.
DOWN - Bangor 1837 [WT]; Newcastle 1836 & 1851 [WT]; Annalong 1837 [WT] & 1982 [*Sublit Surv*].
ANTRIM - Portrush 1856 [WS]; Brown's B [C].
LONDONDERRY - Dredged off the Roe Br & at Magilligan c.1948 (HB).

DICTYOTALES

Dictyopteris Lamour.

Dictyopteris membranacea (Stackh.) Batt.
(Pl.10) **-A-**
Unknown in the littoral but rare in the deep sublittoral of NI to depths of c.25m on exposed shores. A southern species but in NI known mainly from Rathlin I.
ANTRIM - Rathlin I: Farganlack Pt, Altachuile B, W of Derginan Pt & Bull Pt 1984 & 1985, W Church B 1985 (*Sublit Surv*), Black Hd 1985 [*Sublit Surv*], White Cliffs 1984 & 1985 (*Sublit Surv*) & 1984 [*Sublit Surv*], Arkill B 1984 [*Sublit Surv*]; Broad Sound 1982 (*Sublit Surv*); Ringfad Pt 1983 [*Sublit Surv*].

Dictyota Lamour.

Dictyota dichotoma (Huds.) Lamour. **DAL**
Common and locally abundant. Low littoral rock pools and sublittoral. First recorded in NI "From the shore of C.

Down July 1806" [JT].
DOWN, ANTRIM and LONDONDERRY - Common.

var. **latifrons** Holm. et Batt.   **—L**
LONDONDERRY - Cast up at Magilligan pre-1948 (HB).

Taonia J. Ag.

Taonia atomaria (Woodw.) J. Ag.   **D—**
Very rare indeed; only recorded recently in NI. A southern species in the British Is.
DOWN - N of Greenore Pt sublittoral 1983 [*Sublit Surv*] F4922, F4325.

SPOROCHNALES

Carpomitra Kütz.

Carpomitra costata (Stackh.) Batt.   **-A-**
Until recently unknown from NI. Rare, recorded sublittorally on exposed coasts to depths of 25m in Co Antrim (as well as in Co Donegal) but not yet from Cos Londonderry or Down.
ANTRIM - Altachuile B 1983 & 1985 [*Sublit Surv*], Bull Pt 1985 (*Sublit Surv*) & at Arkill B [*Sublit Surv*], all on the E coast of Rathlin I; Kinbane Hd 1985 [*Sublit Surv*]; Murlough B 1984 [*Sublit Surv*]; Torr Hd 1983 [*Sublit Surv*]; Ringfad Pt 1983 [*Sublit Surv*].

Sporochnus C. Ag.

Sporochnus pedunculatus (Huds.) C. Ag.
                                        DAL
A rare species with few records from NI. Sublittoral to depths of 25m.
DOWN - Bangor 1835 [WT] (several publications refer to this find); Audley's Pt 1983 [*Sublit Surv*]; Killowen Bank & Greenore Pt in Carlingford L 1983 [*Sublit Surv*].
ANTRIM - Cooraghy B 1985 [*Sublit Surv*]; dredged from N Belfast L 1896 (TJ

& RH 1896).
LONDONDERRY - Mouth of R Bann
among rejectamenta (DM in *Fl Hib*);
Magilligan dredged pre-1948 (HB 1951).

DESMARESTIALES

Desmarestia Lamour.

Desmarestia aculeata (L.) Lamour.　**DAL**
Common on rocks in the low and
sublittoral. Often found twisted around
the stipes of *Laminaria* or cast ashore
after a storm.
First recorded from NI in Cairnlough B
1836 (JSD 1837).
DOWN, ANTRIM and
LONDONDERRY - Common.

D. dresnayi Lamour. ex Leman　　**-A-**
Only found in the sublittoral and until
recently only from Devon, Cornwall and
Argyll in Britain, and Kerry and Donegal
in Ireland (*Sea B I*).
ANTRIM - Altacorry Hd Rathlin I 1985
[*Sublit Surv*]. This is the first record for
NI.

D. ligulata (Lightf.) Lamour.　　**DAL**
Rather rare in the littoral, but more
common in the sublittoral to a depth of
about 17m. Apparently more common in
Co Antrim than Co Down.
DOWN - Bangor 1835 (JLD in WT
1836); Donaghadee Sound 1985 (*Sublit
Surv*); Ringhaddy Sound 1985 (*Sublit
Surv*); Portaferry 1950 [WRM]; near
Arno's Vale c.1931 (JMW 1931).
ANTRIM - Rathlin I 1983, 1984 & 1985
(*Sublit Surv*); Little Skerries 1982 (*Sublit
Surv*); Portrush 1898 [HH] & 1984 [*Sublit
Surv*]; North B Portrush 1915 [SW];
White Rocks c.1933 (WSW & JWW
1933b) & 1980 [OM]; Portballintrae 1960
[MPHK]; Giant's Causeway pre-1836 (*Fl
Hib*) & pre-1978 (*Causeway Proj*), near
Giant's Causeway 1980 [OM] & N of
Giant's Causeway 1982 (*Sublit Surv*);

White Pk B 1915 [MWR]; off Larne drift
1838 [GCH?].
LONDONDERRY - Portstewart 1949
[WRM] & Portstewart Pt 1984 (*Sublit
Surv*); Rinagree Pt 1984 (*Sublit Surv*).

D. viridis O.F. Müll.　　　**DAL**
A rather rare species in the low littoral,
but much more common in the sublittoral
to a depth of 25m.
DOWN and ANTRIM - Common in the
sublittoral [*Sublit Surv*].
LONDONDERRY - Bann Estuary 1982
[*Sublit Surv*]; N W Portstewart Pt 1984
(*Sublit Surv*); N W Rinagree Pt 1984
(*Sublit Surv*).

Arthrocladia Duby

Arthrocladia villosa (Huds.) Duby　**DA-**
Rare, low littoral and sublittoral down to
at least 19m.
ANTRIM/DOWN - Belfast L pre-1872
(*BA* 1874).
DOWN - "Co Down" pre-1898 (RLIP) &
pre-1902 (Mrs Gatty in *BA* 1902); S of
Killowen Bank, Carlingford L 1983
[*Sublit Surv*].
ANTRIM - E coast & Church B Rathlin I
1985 [*Sublit Surv*]; Loughan B 1983
[*Sublit Surv*]; N of Red B 1983 [*Sublit
Surv*]; Carrickfergus pre-1846 (WMcC in
Phycol Brit); N side of Belfast L dredged
1896 (TJ & RH 1896).

DICTYOSIPHONALES

Myriotrichia Harv.

Myriotrichia clavaeformis Harv.　**DA-**
(Incl *M. filiformis* Harv.)
Rarely recorded but probably overlooked.
Small filamentous tufts, epiphytic on
various algae and *Zostera*.
DOWN - Herring B 1985 (*NILS*);
Kircubbin Pt 1985 (*NILS*); Black Neb
1985 (*NILS*); Horse I 1986 (*NILS*);
Yellow Rocks 1986 (*NILS*); Strangford L

narrows 1838 [WT]; St John's Pt 1985
(*NILS*); Newcastle 1838 [WT];
Glasdrumman Port 1985 (*NILS*).
ANTRIM - N of Ireland pre-1847 (*Phycol
Brit*); Belfast & N of Ireland generally
(*Batters' Cat*); Port Vinegar 1985 (*NILS*);
Red Arch 1985 (*NILS*).

Isthmoplea Kjellm.

Isthmoplea sphaerophora (Carm. ex
Harv.) Kjellm.   **DA-**
Until recently this small filamentous alga
was considered very rare as there were
only two records of it in NI.
First recorded at Newcastle in 1838 (see
below).
DOWN - Foreland Pt 1986 (*NILS*);
Whitechurch 1987 (*NILS*); Kircubbin Pt
1985 (*NILS*); Robin's Rock 1986 (*NILS*);
Castle I Pt 1986 (*NILS*); The Dorn 1986
(*NILS*); Ringburr Pt 1985 (*NILS*);
Limestone Rock 1986 (*NILS*); Chapel I
causeway 1986 (*NILS*); Granagh B 1986
(*NILS*); Kilclief Pt 1986 (*NILS*);
Carrstown Pt 1986 (*NILS*); Ballyquintin
Pt 1986 (*NILS*); Mill Quarter B 1985
(*NILS*); Sheepland Harbour 1985 (*NILS*);
Rathmullan Pt 1985 (*NILS*); Craigalea
1980's (*NILS*); St John's Pt 1985 (*NILS*);
Bloody Br 1985 (*NILS*); Newcastle 1838
[WT]; Wreck Port 1985 (*NILS*); S of
Nicholson's Pt 1985 (*NILS*).
ANTRIM - Cushendun S 1985 (*NILS*);
Garron Harbour 1985 (*NILS*); The
Maiden Rocks c.1869 (GD 1871); Black
Hd E 1986 (*NILS*).

Stictyosiphon Kütz.

Stictyosiphon griffithsianus (Le Jol.)
Holm. et Batt.   **-A-**
A very rare species although possibly
overlooked. Only one old record.
ANTRIM - Epiphytic on a young *Fucus
serratus* at The Gobbins 1898 (HH 1899).

Striaria Grev.

Striaria attenuata (Grev.) Grev.   **D—**
Very rare. Few old records only.
Sublittoral.
DOWN - Bangor 1831 [JLD]; White
Rock drift c.1934 (MJL 1939); Portaferry
dredged from sublittoral 1838 [WT];
Strangford L pre-1841 (WT in *Harvey's
Man*). This record may refer to the
Portaferry specimen mentioned above.

Asperococcus Lamour.

Asperococcus compressus Griff. ex Hook.
                                   DA-
Rare, unknown from NI until recently.
Littoral and sublittoral.
All the records below are from the *NILS*.
DOWN - South I 1986; Herring B 1985;
Mahee I 1985; Ringhaddy Rapids 1987;
Mill Quarter B 1985; Coney I 1985;
Rathmullan Pt 1985.
ANTRIM - On Rathlin I at E Lighthouse
platform 1985; The Burnfoot 1984; The
Ladle 1985; Larry Bane B 1984.

A. fistulosus (Huds.) Hook.   **DAL**
Common as an epiphyte in pools of the
mid and low littoral and sublittoral.
DOWN and ANTRIM - Common.
LONDONDERRY - Balls Pt 1987
(*NILS*); Portstewart 1949 [WRM];
Rinagree Pt 1984 (*NILS*).

A. turneri (Sm.) Hook.   **DAL**
Rather rare but said by some to be locally
abundant. Recorded from rock pools of
the littoral into the sublittoral.
DOWN - Holywood [anon]; Strangford L
pre-1841 (WT); NE Mahee I 1983 [*Sublit
Surv*]; E of Roe I 1983 [*Sublit Surv*];
Yellow Rocks 1986 (*NILS*); Killinchy
1837 [WT]; Portaferry 1837 [WT];
Ardglass 1851 [WT]; Killough 1898
[HH].
ANTRIM - Nearest coast ... to The
Maiden Rocks c.1871 (GD 1871).

LONDONDERRY - Dredged off
Magilligan 1937-1948 (HB 1951).

Litosiphon Harv.

Litosiphon laminariae (Lyngb.) Harv. **DA-**
(Incl *L. pusillus* (Carm. ex Hook.) Harv.)
A small epiphytic alga forming tufts up to
40mm long mainly on *Chorda* and *Alaria*.
Low littoral and upper sublittoral.
DOWN - Coast of Down pre-1841 (WT
in *Harvey's Man*); Bangor 1835 [WT];
Sandeel B 1972 [OM]; Wallaces Rocks
1984 (*NILS*); Mid I B 1986 (*NILS*);
Taggart I 1986 (*NILS*); Carrstown Pt 1986
(*NILS*); Sheepland Harbour 1985 (*NILS*);
Coney I 1985 (*NILS*); Killough 1905 (JA
1913); Wreck Port 1985 (*NILS*); S of
Nicholson's Pt 1985 (*NILS*).
ANTRIM - Antrim pre-1841 (DM in
Harvey's Man); Rathlin I at Stackamore
1987 & E Lighthouse platform 1985
(*NILS*); Peggy's Hole 1985 (*NILS*); Port
Gorm 1985 (*NILS*); The Ladle 1985
(*NILS*); Port-na-Tober & Port-na-Tober
Hd 1987 (*NILS*); Cushendun 1985 (*NILS*);
Cushendall prob pre-1866 [prob WHH];
White Lady 1985 (*NILS*); Garron Harbour
1985 (*NILS*); Ballygalley 1975 [OM];
Seacourt 1985 (*NILS*); The Gobbins 1898
[HH].

Punctaria Grev.

Punctaria latifolia Grev. **DA-**
Rare with few recently confirmed records.
P. plantaginea is included by some (*N A
Checklist*) as a cold-water form, however
this requires further investigation (*Sea B
I*). As the *NILS* follows the *N A Checklist*,
some records of *P. plantaginea* may be
included below.
ANTRIM/DOWN - Near Belfast pre-
1830 (JLD in *Alg Brit*), probably it is this
record which is reported in several of the
old references (incl *Harvey's Man, Fl
Hib, Seaw An Co* & *Phycol Brit*).
DOWN - Sandeel B 1973 [OM det HB];

Ballywalter 1836 [WT]; Castle I Pt 1986
(*NILS*); The Dorn 1986 (*NILS*);
Ballyhenry Pt 1975 [OM det HB];
Kearney 1985 (*NILS*); Chapel I causeway
1986 (*NILS*); Bloody Br 1985 (*NILS*).
ANTRIM - The Ladle 1985 (*NILS*); Port-
na-Tober 1987 (*NILS*); nearest coast to
The Maiden Rocks c.1871 (GD 1871).

P. plantaginea (Roth) Grev. **DA-**
Very rare. No recent records at all from
NI. Included within the synonymy of *P.
latifolia* (*N A Checklist*), however, RLF
(*Sea B I*) considers this requires further
investigation. As the *NILS* follows the
nomenclature of *N A Checklist* some *P.
plantaginea* records may be included
under *P. latifolia*.
ANTRIM/DOWN - Near Belfast pre-
1830 (JLD in *Alg Brit*) & Belfast pre-
1857 (JLD in *Brit Seaweeds*) these may,
or may not, be based on the same find.
DOWN - Bangor 1836 [WT]; Ballywalter
1836 [WT].
ANTRIM - Belfast pre-1902 (*Batters'
Cat*).

P. tenuissima (C. Ag.) Grev. **DA-**
A rare species for which there are few
records for all Ireland. Possibly
conspecific with *P. latifolia*. Epiphytic in
the low littoral and upper sublittoral.
DOWN - Drummond I 1986 (*NILS*);
Marlfield Rocks 1985 (*NILS*); Holm B
1984 (*NILS*); S of Newcastle 1976 [OM
conf HB]; Bloody Br 1985 (*NILS*); Wreck
Port 1985 (*NILS*); S of Nicholson's Pt
1985 (*NILS*).
ANTRIM - Rathlin I at Black Hd 1986
(*NILS*); Port Gorm 1985 (*NILS*); The
Ladle 1985 (*NILS*); White Lady 1985
(*NILS*).

Pogotrichum Reinke

Pogotrichum filiforme Reinke **DA-**
Small epiphytic alga forming tufts on
various species especially *Laminaria*. Not

previously recorded from NI.
All the records below from the *NILS*.
DOWN - Chapel I causeway 1986;
Kearney 1985; Craigalea 1980's; St
John's Pt 1985; Glasdrumman Port 1985.
ANTRIM - Carrickarade 1986; Garron
Harbour 1985; Campbeltown 1986;
Straidkilly Pt 1985.

Dictyosiphon Grev.

Dictyosiphon chordaria Aresch.   **DA-**
Very rare. All records are recent and from
the *NILS*.
DOWN - Orlock Pt 1985; Barkley Rocks
1984; Black Neb 1985; Kearney 1985;
Ardglass B 1985; Rathmullan 1985;
Wreck Port 1985; Greencastle Rocks
1985.
ANTRIM - Port Gorm 1985.

D. foeniculaceus (Huds.) Grev.   **DA-**
Rarely recorded in NI although more
common in Co Down than Co Antrim.
DOWN/ANTRIM - Near Belfast pre-
1830 (JLD in *Alg Brit*).
DOWN - Common.
ANTRIM - Portbraddan drift 1983 (OM);
Cushendall 1830 [WHH]; Straidkilly Pt
1985 (*NILS*); Larne pre-1853 [JLD];
Belfast pre-1902 (*Batters' Cat*).

Colpomenia (Endlicher) Derbès et Solier

Colpomenia peregrina (Sauv.) Hamel
Oyster Thief   **DAL**
(*C. sinuosa* (Roth) Derbès et Solier)
An infrequent species, probably native to
the Pacific from whence it has spread to
the N Atlantic. Epiphytic on various algae
in the littoral.
It was first recorded in England in 1908
and in Strangford L in 1934 (MJL 1935b).
DOWN - Sandeel B 1972 [OM]; near
Donaghadee 1984 [OM]; opposite
Mountstewart 1934 (MJL 1935b); Yellow
Rocks 1986 (*NILS*); Ardkeen 1976 [OM

det HB]; Marlfield B 1934 (MJL 1935b);
Ballyhenry B 1934 (MJL 1935b);
Kearney 1977 & 1987 [OM]; Killough
1936 (HB 1937); Annalong pre-1952
(MJL 1952); Kilkeel pre-1952 (MJL
1952).
ANTRIM - Portballintrae pre-1935 (MJL
1935b), 1961 [MPHK]; between
Ballycarry & Magheramorne in Larne L
drift 1935 (MJL 1935b); Hood's Ferry I
Magee drift 1935 (HB 1937).
LONDONDERRY - Portstewart 1935
(HB 1937).

Petalonia Derbès et Solier

Petalonia fascia (O.F. Müll.) Kuntze   **DAL**
Uncommon in the eulittoral. Laboratory
cultures suggest at least two phases are
involved in the life-history of *P. fascia*.
One is the erect bladed plant considered
here, the other is an encrusting phase,
possibly *Stragularia clavata*. However
until the situation is clarified it is
considered best to treat them here as
separate species.
DOWN and ANTRIM - Common.
LONDONDERRY - Rinagree B 1983
[OM] & Rinagree Pt 1984 (*NILS*).

P. filiformis (Batt.) Kuntze   **D—**
Small flat alga up to 8cm long, epilithic.
Not recorded in Ireland previous to the
NILS (MDG 1978).
DOWN - Ballyhenry B 1986 (*NILS*).

P. zosterifolia (Reinke) Kuntze   **DA-**
A rare epilithic alga growing in tufts in
the littoral. Only one previous record
from Ireland (MDG 1978).
Only recorded in NI by the *NILS*.
DOWN - Ballymacormick Pt 1984;
Mahee I 1985; Rathmullan Pt 1985;
Wreck Port 1985.
ANTRIM - On Rathlin I at The Dutchman
1985; Gt Stookan 1986; Port Moon B
1986; Dunineny Castle 1987; Murlough B
1985; Portmore 1980; Whitebay Pt 1986;

Halfway House 1985.

Scytosiphon C. Ag.

Scytosiphon lomentaria (Lyngb.) Link
DAL
A common species of the littoral and upper sublittoral. Further work is required on the life-history of this species as other taxa appear to be involved.
First recorded at Donaghadee pre-1825 [JT] F306.
DOWN and ANTRIM - Common.
LONDONDERRY - Black Rock 1917 [SW]; Rinagree B 1983 [OM] & Rinagree Pt 1984 (*NILS*).

LAMINARIALES

Chorda Stackh.

Chorda filum (L.) Stackh.   **DAL**
Sea Lace
Common and locally abundant in more sheltered sites near low water and sublittoral to 25m.
First NI record from Carnlough B 1833 [WT].
DOWN, ANTRIM and LONDONDERRY - Common.

C. tomentosa Lyngb.   **DAL**
A very rare species of the low littoral and sublittoral. Rarely recorded but possibly overlooked.
DOWN - Ballymacormick Pt 1984 (*NILS*); Wreck Port 1985 (*NILS*).
ANTRIM - "Coast of Antrim..." pre-1902 (*Batters' Cat*); W Lighthouse Rathlin I 1983 [*Sublit Surv*]; Garron Harbour 1985 (*NILS*).
LONDONDERRY - Rinagree Pt 1984 (*NILS*).

Laminaria Lamour.

Laminaria digitata (Huds.) Lamour. **DAL**
Oarweed or Tangle

The dominant alga on all shores where there is rock, at and just below, low littoral.
DOWN, ANTRIM and LONDONDERRY - Very common.

L. hyperborea (Gunn.) Fosl.   **DAL**
Curvie
A common and dominant species in the upper sublittoral.
DOWN, ANTRIM and LONDONDERRY - Common.

L. longicruris Pyl.   **-A-**
One or two drift specimens referred to under this name have been found in Co Antrim.
ANTRIM - Near Dunluce Castle 1850 (WHH in *Phycol Brit*); "a battered plant was picked up by Dr W H Harvey, at the Giant's Causeway" 1853 (*Gifford's Mar Bot*). Probably these two records refer to one and the same find.

L. saccharina (L.) Lamour.   **DAL**
Sea Belt, Sugar Kelp or Sugarwrack
Frequent in the low littoral and upper sublittoral all around the three counties wherever there is rock or pebbles.
DOWN and ANTRIM - Frequent.
LONDONDERRY - Balls Pt 1987 (*NILS*); Black Rock pre-1836 (DM in *Fl Hib*); Rinagree Pt 1984 (*NILS*).

Saccorhiza Pyl.

Saccorhiza polyschides (Lightf.) Batt.
Furbelows   **DAL**
A common low littoral and sublittoral species down to 18m, well known in the drift.
First recorded in NI at Larne c.1830 (*Alg Brit*).
DOWN and ANTRIM - Common.
LONDONDERRY - Only one record: near Rinagree Pt 1983 (OM).

Alaria Grev.

Alaria esculenta (L.) Grev.   **DAL**
Dabberlocks
   Widespread and common species of low
   littoral and sublittoral to 25m especially in
   exposed conditions. Rare in the shelter of
   Strangford L. First recorded in NI as
   "*Fucus esculentus*" from Glenarm by JT
   in July 1808 (MPHK 1966).
   DOWN, ANTRIM and
   LONDONDERRY - Common.

FUCALES

Ascophyllum Stack.

Ascophyllum nodosum (L.) Le Jol.   **DAL**
Knotted Wrack or Egg Wrack   (Pl.5)
   Widespread and abundant especially in
   sheltered locations. Mid-littoral.
   DOWN, ANTRIM and
   LONDONDERRY - Widespread and
   abundant.
   There is a lot of variation in *Ascophyllum*,
   some of which is genetic and some
   phenotypic (GR- pers comm). As a result
   there is a lack of agreement as to whether
   there are subspecies, varieties or ecads.

var. **minor** Turn.   **-A-**
   Rare. Attached to small stones
   embedded in the mud at high water
   or free-living embedded in mud.
   ANTRIM - Near Magheramorne
   1937 (MJL 1960b) & at a number
   of sites between Oldmill B &
   Ballycarry railway station in Larne
   L 1937 (MJL 1949 & 1960a).

ecad **scorpioides** (Hornem.) Hauck **DA-**
   Rare. Occurs attached, as loose-
   lying clumps or as a turf in the mud
   of a marsh.
   DOWN - In brackish water at
   Kircubbin B 1989 [OM conf GR];
   Yellow Rocks & Ballyhenry I both
   c.1935 (MJL 1935a); Dundrum

Inner B 1985 [OM].
ANTRIM - Near Newmill in Larne
L 1989 [OM conf GR].

ecad **mackaii** Cotton   **D—**
   Mackay's Wrack
   Generally rare, but not uncommon
   in certain places. Free growing.
   DOWN - White Rock 1936 (MJL
   1937); Yellow Rocks c.1935 (MJL
   1935a); Ballyhenry I c.1935 (MJL
   1935a) still there 1975 [OM].

Fucus L.

Fucus ceranoides L.   **DAL**
Horned Wrack
   Not uncommon where fresh water enters
   the sea in sheltered localities. Littoral.
   DOWN/ANTRIM - Shores of the Lagan
   pre-1825 [JT]; near Belfast pre-1833 (*Eng
   Fl*) & Belfast L pre-1836 (*Fl Hib*).
   DOWN - Ballymacormick Pt 1984
   (*NILS*); near Donaghadee pre-1825 [JT];
   tributary of Comber R 1977 [OM]; Rolly
   I 1973 [OM det GR]; Ringhaddy Rapids
   1987 (*NILS*); Kircubbin 1990 [OM]; near
   Saltwater Br 1989 [OM conf GR];
   Portavogie 1915 [MWR]; Ardkeen 1975
   (AH & RS 1976); Marlfield Rocks 1985
   (*NILS*); Black Causeway 1986 (*NILS*);
   Mill Quarter B 1985 (*NILS*); Ardglass
   c.1932 (MJL & JMcG 1932); near
   Killough c.1934 (MJL & JMcG 1934);
   near St John's Pt 1988 [OM]; Dundrum
   Inner B 1984 & 1985 [OM]; Newcastle
   pre-1825 [JT] & pre-1852 [WT]; Mill B
   1981 [OM det GR] & 1980's (*NILS*).
   ANTRIM - Portballintrae 1960 [MPHK];
   Bush R outlet 1960 [MPHK]; Cushendun
   1984 [OM]; mouth of Glenariff R 1981
   [PH]; Turnley's Port 1985 (*NILS*); Larne
   c.1830 (JLD in *Alg Brit*), pre-1852 [JLD],
   pre-1902 (*Batters' Cat*) also Larne L
   c.1937 (MJL 1960a, & 1961b) & Newmill
   in Larne L 1989 [OM conf GR].
   LONDONDERRY - New Br 1987
   (*NILS*); Culkeeragh c.1948 (HB 1951);

Lower Doaghs 1984 (*NILS*); Bann
Estuary 1982 [*Sublit Surv*].

f. **ramosissima** Lynn   D—
   Very rare. This was described as a
   new form in 1935 from a brackish
   marsh.
   DOWN - Ringbane c.1935 (MJL
   1935b).

var. **glomerata** (Batt.) Lynn   D—
   Very rare.
   DOWN - Near White Rock where
   the Ganaway Burn flows over the
   shingle 1934 (MJL 1935b).

F. cottonii Wynne et Magne   D—
(*F. muscoides* (Cotton) Feldm. et Magne)
(*F. vesiculosus* var. *muscoides* Cotton)
(*F. vesiculosus* var. *balticus sensu* P. et H.
Crouan)
(non *F. balticus* C. Ag.)
   A rare species of salt marshes in brackish
   conditions forming a "mossy" growth on
   earth banks in the supralittoral close to
   vascular plants. Wynne & Magne (1991)
   explained the confusion of this entity with
   F. balticus C. Ag. and proposed the new
   name as above.
   DOWN- Rolly I 1975 [OM] F408, 1979
   [OM conf GR as *F. muscoides*] F2337 &
   1985 [OM] F4703; three sites in
   Strangford L: Horse I where the plant
   forms a "turf" mixed with *Glyceria &
   Salicornia*, Ballywallon marsh and at
   Castle I c.1935 (MJL 1935a); Dundrum B
   1902 [CHW]; Mill B Carlingford L 1981
   [OM conf M.J. Wynne & F. Magne]
   F2589 duplicate.

F. serratus L.   DAL
Serrated or Toothed Wrack
   Widespread and abundant in the lower
   littoral except in the most exposed
   positions.
   DOWN, ANTRIM and
   LONDONDERRY - Abundant.

f. **abbreviatus** Kjellm.   D—
   Very rare. Epilithic in pools at
   mid-littoral.
   DOWN - Kilclief c.1934 (MJL
   1935b).

F. spiralis L.   DAL
Spiral or Twisted Wrack
   Widespread and common except in very
   exposed positions, forming a zone along
   the shore in the upper littoral.
   DOWN, ANTRIM and
   LONDONDERRY - Common.

There is some disagreement about the status
of the varieties described in the literature.
Those from NI include:-

var. **nanus** Stackh.   DAL
   A rare dwarf plant of salt-marshes
   as well as rocky coasts in very
   exposed positions.
   DOWN - Salt-marshes of Strangford L
   at Yellow Rocks c.1935 &
   Quarterland B
   c.1935 (both MJL 1935b).
   ANTRIM - Marsh near Ballycarry in
   Larne L c.1935 (MJL 1935b) &1937
   (MJL 1960a).
   LONDONDERRY - Eglinton &
   Culkeeragh c.1951 (HB 1951).

var. **platycarpus** (Thur.) Børg.   D—
   Rare, few records for all Ireland.
   DOWN   -
   Sketrick I c.1936 (MJL 1937);
   betweenWarrenpoint & Rostrevor
   c.1931(JMW 1931) & Rostrevor c.
   1932 (JMW 1932).

F. vesiculosus L.   DAL
Bladder Wrack
   Widespread and abundant, except in
   exposed positions. Forming a zone along
   the shore in the mid-littoral.
   DOWN, ANTRIM and
   LONDONDERRY - Abundant.

Several varieties of this species have been described, among them :-

var. **angustifolius** L.   —L
Very rare. Few records for all Ireland.
LONDONDERRY - Culkeeragh with *F. vesiculosus* c.1948 (HB 1951).

var. **linearis** (Huds.) Powell   —L
Very rare. Only recorded on exposed shores.
LONDONDERRY - Rinagree Pt 1984 (*NILS*).

var. **volubilis** Turn.**D—**
Very rare, few records for all Ireland. Only recorded from salt-marshes.
DOWN - Strangford L several sites: Castle Espie c.1935 & Ringneill Br c.1935 (MJL 1935a); Rolly I 1978 [OM conf GR]; Ballywallon c.1935 (MJL 1935a).

Pelvetia Dcne. et Thur.

Pelvetia canaliculata (L.) Dcne. et Thur.
Channelled Wrack   **DAL**
Widespread and common on all but the most exposed shores. Forms a zone along the shore at the upper littoral. Usually on rock but occasionally elsewhere, in salt-marsh conditions it has been found growing on *Halimione portulacoides* (Chenopodiaceae) [OM] and even an old vehicle tyre (OM).
DOWN, ANTRIM and LONDONDERRY - Common & abundant.

var. **libra** Baker   **D—**
Very rare, few records for all Ireland. A free growing variety.
DOWN - In Strangford L on salt-marsh opposite the point where the Ganaway Burn enters the beach c.1936 (MJL 1937); Ballymorran B

c.1935, Yellow Rocks c.1935 & Ballyhenry B c.1935 (all MJL 1935b).

Himanthalia Lyngb.

Himanthalia elongata (L.) S.F. Gray **DAL**
Sea Thongs   (Pl. 6)
Very common on rocky shores, especially relatively exposed ones, low littoral.
DOWN and ANTRIM - Common.
LONDONDERRY - Rinagree Pt 1984 (*NILS*).

Bifurcaria Stackh.

Bifurcaria bifurcata Ross   **-A-**
In 1804 WW published a catalogue of Irish plants: *Plantae rariores in Hibernia Inventae*. On page 158 he recorded *Bifurcaria bifurcata*: "found closely adhering to the rocks by rather a broad base at the Giant's Causeway; and if I do not mistake, on top of the basaltic columns in the sea, immediately at the end of the causeway." (de Valéra 1962). If Wade identified this specimen correctly it is the first published record for Ireland of this species and the only one for NI. However there is no voucher specimen and it would be unwise to place too much confidence in the record. It is a southern species common in N Co Donegal.

Cystoseira C. Ag.

Cystoseira baccata (Gmel.) Silva   **-AL**
This southern species is limited in its distribution and is rare in NI with no recent records. Sublittoral or intertidal pools.
ANTRIM - Shores of Antrim pre-1825 (JT in *Harvey's Man*); Portrush c.1850's [WS det MR] F156. This may be the specimen published under *C. fibrosa* 1854 (WS 1854).
LONDONDERRY - Black Rock pre-1836 (DM in *Fl Hib*).

C. nodicaulis (With.) Roberts　　**DAL**
Usually sublittoral, but also found in rock pools. Rather rare.
DOWN - Near Ballywalter 1806 [JT]; Ardkeen 1976 [OM det MR]; Marlfield Rocks 1985 (*NILS*); Portaferry 1837 [WT]; Quintin B 1952 [WRM]; Granagh B 1976 [OM conf MR]; Ardglass 1838 & 1851 [WT].
ANTRIM - Portrush c.1874 (*BA* 1874); Larne 1836 (JSD 1837), Larne L dredged 19th C [C] & 1904 (JA 1913).
LONDONDERRY - Black Rock c.1836 (DM in *Fl Hib*) this may refer to the specimen from Derry in **BEL** pre-1879 [DM]; Magillagan pre-1841 (GCH in *Harvey's Man*).

C. tamariscifolia (Huds.) Papenf.　　**DA-**
Extremely rare in NI. The two records from Portrush may be based on a single find.
DOWN - Chapel I causeway 1986 (*NILS*).
ANTRIM - Portrush pre-1841 (*Harvey's Man*) & pre-1853 (*Gifford's Mar Bot*); Portballintrae c.1953 (HB & NFM 1953).

Halidrys Lyngb.

Halidrys siliquosa (L.) Lyngb.　　**DAL**
Sea Oak
A common species of rocky shores in low littoral rock pools and the sublittoral to depths of about 20m.
DOWN and ANTRIM - Common.
LONDONDERRY - Few records but probably common: Portstewart 1981 (OM); Rinagree B 1983 (OM) & Rinagree Pt 1984 (*NILS*).

Rhodophyta

Bangiophyceae

PORPHYRIDIALES

Stylonema Reinsch

Stylonema alsidii (Zanard.) Drew   **DA-**
(*Goniotrichum alsidii* (Zanard.) Howe)
   Very rare indeed. Until recently only one
   record repeated in a number of
   publications.
   DOWN - Portaferry dredged adhering to
   Gracilaria verrucosa 1838 (WT det
   WHH in *Phycol Brit* as *Bangia elegans*
   Chauv.); Bar Hall B 1986 (*NILS*).
   ANTRIM - Barney's Pt 1986 (*NILS*).

BANGIALES

Erythrocladia Rosenv.

Erythrocladia irregularis Rosenv.   **DAL**
   Minute filamentous alga more, or less
   disc shaped. Epizoic. Only recorded once
   in Ireland, and never in NI before *NILS*.
   Obviously commonly overlooked due to
   small size.
   All the records below are from the *NILS*.
   DOWN - Common.
   ANTRIM - Rathlin I at: Maria Isla 1985,
   E Lighthouse platform 1985, Doon B
   1985 & Black Hd 1986; The Burnfoot
   1984; Port Gorm 1985; The Ladle 1985;
   Port Moon 1984; Larry Bane B 1984;
   Portdoo 1986; Murlough B 1985; White

Lady 1985; Ringfad 1984; Straidkilly Pt
1985; Drumnagreagh Port 1984; Halfway
House 1985; Riding Stone 1986;
Loughshore Pk 1986.
LONDONDERRY - Rinagree Pt 1984.

Erythrotrichia Aresch.

Erythrotrichia carnea (Dillw.) J. Ag. **DA-**
(Incl *Erythrotrichia bertholdii* Batt.)
   Very small monosiphonous fronds, less
   than 1cm long. Probably not uncommon
   but generally overlooked.
   DOWN - Common (*NILS*).
   ANTRIM - Rathlin I pre-1902 (*Batters'
   Cat*); Peggy's Hole 1985 (*NILS*); Port
   Gorm 1985 (*NILS*); Gt Stookan 1986
   (*NILS*); Ballycastle pre-1902 (*Batters'
   Cat*); Layd Church 1984 (*NILS*);
   Campbelton 1986 (*NILS*); Halfway House
   1985 (*NILS*); Ballygalley Hd 1985
   (*NILS*); Barney's Pt 1986 (*NILS*); dredged
   N side of Belfast L 1896 (TJ & RH 1896);
   Loughshore Pk 1986 (*NILS*).

Erythrotrichopeltis Kornm.

Erythrotrichopeltis boryana (Mont.)
Kornm.   **-A-**
(*Erythrotrichia boryana* (Mont.) Berth.)
   Very rare indeed, no recent records at all.
   ANTRIM - Near harbour at Ballycastle
   &/or at Murlough 1897 (det H Murray in
   RSt 1897).

E. ciliaris (Carm. ex Harv.) Kornm.   -A-
(*Erythrotrichia ciliaris* (Carm. ex Harv.)
Thur.)
Very rare indeed, no recent records.
ANTRIM - Dredged near The Maiden
Rocks c.1869 (GD 1871).

Porphyropsis Rosenv.

Porphyropsis coccinea (J. Ag. ex Aresch.)
Rosenv.   DA-
Small alga, less than 4cm in length. In
depths of 25m, however also recently
recorded in the littoral. Possibly
overlooked.
DOWN - Lighthouse I 1984 & near
Copeland I 1985 (*Sublit Surv*); Herring B
1985 (*NILS*); Mahee I 1985 (*NILS*);
Kearney 1985 (*NILS*); Strangford L
narrows 1983 [*Sublit Surv*]; Wreck Port
1985 (*NILS*); S of Nicholson's Pt 1985
(*NILS*); Greencastle Rocks 1985 (*NILS*).
ANTRIM - Rathlin I 1985 (*Sublit Surv*)
& Stackamore 1987 (*NILS*), E
Lighthouse platform 1987 (*NILS*);
Portawillin 1986 (*NILS*); The Burnfoot
1984 (*NILS*); Port Gorm 1985 (*NILS*);
The Ladle 1985 (*NILS*); Kinbane Hd 1984
(*Sublit Surv*); Torr Hd 1983 [*Sublit Surv*];
Murlough B 1985 (*NILS*); Cushendun
1985 (*NILS*); Ringfad 1984 (*NILS*);
Carnlough B 1983 [*Sublit Surv*];
Drumnagreagh Port 1983 [*Sublit Surv*];
Halfway House 1985 (*NILS*); Seacourt
1985 (*NILS*); Larne 1898 [HH]; near
Curran Pt 1984 (*Sublit Surv*).

Bangia Lyngb.

Bangia atropurpurea (Roth) C. Ag.   DAL
(*B. fuscopurpurea* Lyngb.)
Unbranched filaments up to 10 or 15cm
long on upper rocky shores. Widely
distributed.
DOWN - Bangor B 1930 (MC 1930);
Herring B 1985 (*NILS*); Black Neb inlet
1986 (*NILS*); Ringburr Pt 1985 (*NILS*);
Limestone Rock 1986 (*NILS*); Ballyhenry

B 1986 (*NILS*); Green I 1986 (*NILS*);
Sheepland Harbour 1985 (*NILS*); S of
Nicholson's Pt 1985 (*NILS*); Greencastle
Rocks 1985 (*NILS*).
ANTRIM - Common.
LONDONDERRY - Rinagree Pt 1984
(*NILS*).

Porphyra C. Ag.

Porphyra leucosticta Thur.   DA-
Common but until recently under-
recorded partly because of difficulty of
identification. Epiphytic and epilithic in
littoral and sublittoral.
DOWN - Culloden Hotel 1986 (*NILS*);
Rockport 1979 [OM det PSD]; Swineley
Pt 1986 (*NILS*); Orlock Pt 1985 (*NILS*);
Lighthouse I 1975 [OM det PSD];
Foreland Pt 1986 (*NILS*); Drummond I
1986 (*NILS*); Kearney 1985 (*NILS*);
Down Rock 1986 (*NILS*); Sheepland
Harbour 1985 (*NILS*); Coney I 1985
(*NILS*); St John's Pt 1985 (*NILS*); Wreck
Port 1985 (*NILS*).
ANTRIM - Common.

P. linearis Grev.   DAL
Apparently more common in Co Antrim
than Co Down. Probably under-recorded
being small and found only in the winter
and spring.
ANTRIM/DOWN - Belfast L pre-1902
(*Batters' Cat*).
DOWN - S of Nicholson's Pt 1985
(*NILS*).
ANTRIM - Rathlin I: Portandoon 1986
(*NILS*), Stackamore 1987 (*NILS*), E
Lighthouse platform 1985 (*NILS*), Doon
B 1985 (*NILS*), The Dutchman 1985
(*NILS*), Killeany 1980's (*NILS*); The
Burnfoot 1984 (*NILS*); Peggy's Hole
1985 (*NILS*); The Ladle 1985 (*NILS*);
Port-na-Tober 1987 (*NILS*); Larry Bane
Hd 1985 (*NILS*); Carrickmore 1985
(*NILS*); Fair Hd W 1985 (*NILS*);
Murlough pre-1902 (*BA* 1902); Old Pier
1984 (*NILS*); Turnley's Port 1985 (*NILS*);

Campbeltown 1986 (*NILS*); Larne pre-1902 (*BA* 1902).
LONDONDERRY - Rinagree Pt 1984 (*NILS*).

P. miniata (C. Ag) C. Ag. **DA-**
A northern species rarely recorded in the littoral. More common in the sublittoral to depths of 15m.
DOWN - Near Lighthouse I 1984 [*Sublit Surv*]; W of Rue Pt 1983 [*Sublit Surv*]; N of Chapel I 1982 (*Sublit Surv*).
ANTRIM - Murlough B 1898 [HH]; Loughan B 1983 [*Sublit Surv*]; Cushendall 1902 (*BA* 1902); N of Red B 1983 [*Sublit Surv*]; Drumnagreagh Port 1983 [*Sublit Surv*]; E of Curran Pt [*Sublit Surv*]; SW of Larne channel 1984 [*Sublit Surv*] & E of Middle Bank in Larne L 1984 (*Sublit Surv*); The Gobbins 1985 (*Sublit Surv*).

P. purpurea (Roth) C. Ag. **DAL**
Not common but probably under-recorded. Littoral.
DOWN - Crawfordsburn beach 1986 (*NILS*); Ballymacormick Pt 1984 (*NILS*); Barkley Rocks 1984 (*NILS*); Wallaces Rocks 1984 (*NILS*); Yellow Rocks 1986 (*NILS*); Ringhaddy Rapids 1987 (*NILS*); Selk Rock 1986 (*NILS*); Ardglass B 1985 (*NILS*); Killough Harbour 1985 (*NILS*); Greencastle Rocks 1985 (*NILS*).
ANTRIM - On Rathlin I at Kinrea 1987 (*NILS*); Peggy's Hole 1985 (*NILS*); The Ladle 1985 (*NILS*); Larry Bane Hd 1985 (*NILS*); near harbour at Ballycastle &/or Murlough 1897 (RSt); Cushendun 1985 (*NILS*); Layd Church 1984 (*NILS*); Ringfad 1984 (*NILS*); Campbeltown 1986 (*NILS*); Drumnagreagh Port 1984 (*NILS*); The Maiden Rocks c.1869 (GD 1871); Riding Stone 1986 (*NILS*).
LONDONDERRY - Rinagree Pt 1984 (*NILS*).

P. umbilicalis (L.) J. Ag. **DAL**
Common and often abundant especially in the upper littoral.
DOWN and ANTRIM - Common.
LONDONDERRY - Lower Doaghs 1984 (*NILS*); Downhill Strand 1984 (*NILS*); Rinagree Pt 1984 (*NILS*).

Florideophyceae

NEMALIALES

Audouinella Bory

Audouinella alariae (Jonss.) Woelk. **-A-**
Minute branched alga with erect filaments to 1mm in length. Epiphytic in the lower littoral. Apparently very rare but probably under-recorded due to its small size and difficulty in identification.
ANTRIM - Kinrea on Rathlin I 1987 (*NILS*); Portrush 1904 (JA 1904a).

A. caespitosa (J. Ag.) Dixon **D—**
Very rare, only one NI record. Sublittoral.
DOWN - W of Rue Pt 1983 (*Sublit Surv*).

A. concrescens (Drew) Dixon **DA-**
Minute filamentous alga with a prostrate base. Epizoic on shells and possibly other hard surfaces usually in the sublittoral.
All the records below are from the *NILS*.
DOWN - Foreland Pt 1986; Herring B 1985; Audley's Castle rocks 1985; Sheepland Harbour 1985; Coney I 1985; Ringboy 1985; Greencastle Rocks 1985.
ANTRIM - Rathlin I in Doon B 1985; Murlough B 1985.

A. daviesii (Dillw.) Woelk. **DA-**
Probably more common than these two records suggest. May be epiphytic on a wide variety of species but most frequently on *Palmaria palmata*.
DOWN - Sandeel B epiphytic on *P. palmata* drift 1973 [OM det MDG in OM 1974].
ANTRIM - The Highlandman's Bonnet B epiphytic on *P. palmata* 1980 [OM conf

FWK].

A. floridula (Dillw.) Woelk.　　　**DAL**
A common species; the basal branches
bind sand particles together forming
patches on the rocks; erect branches up to
30mm long. Very similar to *A. purpurea*.
One of the earliest records from NI is
from "Antrim" pre-1836 (Dr Scott in *Fl
Hib* under the name *Trentepohlia
floridulum* Harv.).
DOWN and ANTRIM - Common.
LONDONDERRY - Lower Doaghs 1984
(*NILS*); Downhill Strand 1984 (*NILS*);
Rinagree Pt 1984 (*NILS*).

A. membranacea (Magn.) Papenf.　　**DA-**
Very rare, endozoic in Hydrozoa and
algae. Minute filaments producing short
erect branches of 3 - 6 cells in length from
a prostrate base.
DOWN - Craiglewey 1984 (*NILS*);
Greencastle Rocks 1985 (*NILS*).
ANTRIM - Drumnagreagh Port 1984
(*NILS*); dredged from the N side of
Belfast L 4 July 1896 [RH] & (TJ & RH
1896).

A. microscopica (Näg.) Woelk.　　**DAL**
Plants microscopic, consisting of a basal
cell and a single erect branching filament
less than 1mm in length. Epiphytic in the
littoral and sublittoral. No doubt under-
recorded due to its minute size.
All the records below are from *NILS*.
DOWN - Swineley Pt 1986;
Ballymacormick Pt 1984; Orlock Pt 1985;
Barkley Rocks 1984; South I 1986;
Herring B 1985; Black Neb 1985; Horse I
1986; Holm B 1984; Granagh B 1986;
Craiglewey 1984; Ringboy 1985.
ANTRIM - Rathlin I at: Portandoon 1986,
Portawillin 1985 & Doon B 1985; The
Burnfoot 1984; Peggy's Hole 1985; Port
Gorm 1985; Gt Stookan 1986;
Carrickarade 1986; Carrickmore 1985;
Portdoo 1986; Murlough B 1985; Layd
Church 1984; Turnly's Port 1985;

Ringfad 1984; Seacourt 1985; Riding
Stone 1986; Chapman's Rock 1987.
LONDONDERRY - Downhill Strand
1984; Rinagree Pt 1984.

A. purpurea (Lightf.) Woelk.　　**DAL**
A common species, very similar to *A.
floridula*, the differences being mainly
microscopic. Erect branches up to 30mm
long. It is not as efficient a sand binder as
A. floridula.
First recorded from NI in a "Limestone
cave on the northern coast of the county
of Antrim, near Ballycastle" 1799 (RB in
Eng Bot).
DOWN and ANTRIM - Common.
LONDONDERRY - New Br 1987
(*NILS*); Rinagree Pt 1984 (*NILS*).

A. secundata (Lyngb.) Dixon　　**DAL**
A much smaller species that either *A.
purpurea* or *A. floridula* with erect axes
no more than 2mm long. Epiphytic on
larger algae and *Zostera*. Until recently
under-recorded.
There is doubt as to whether this is a
species independent from *A. virgatula*.
DOWN and ANTRIM - Common.
LONDONDERRY - Rinagree Pt 1984
(*NILS*).

A. sparsa (Harv.) Dixon　　　　**DA-**
Minute tufted alga with erect axes to
0.5mm. Epiphytic on other algae.
Only recorded in NI by *NILS*.
DOWN - Ballymacormick Pt 1984;
Wallaces Rocks 1984; Mahee I 1985;
Black Neb 1986; The Dorn 1986;
Ringburr Pt 1985; S of Nicholson's Pt
1985.
ANTRIM - The Burnfoot 1984.

A. virgatula (Harv.) Dixon　　　**DA-**
There is doubt as to whether this is a
different species from *A. secundata*.
Erect axes up to 7mm long rise from a
much branched base. Epiphytic on other
algae in the lower littoral and sublittoral.

Apparently rare but may be more common than these records suggest.
DOWN - Ardglass epiphytic on *Porphyra umbilicalis* pre-1852 [WT].
ANTRIM - Oweydoo on Rathlin I 1987 (*NILS*); Antrim coast pre-1902 (*Batters' Cat*); White Pk B 1915 [MWR].

Schmitziella Born. et Batt.

Schmitziella endophloea Born. et Batt.
incertae sedis   **DAL**
Possibly not so rare a species as the records suggest. Rarely recorded but probably overlooked. It grows endophytically between the outer layers of the cell walls of *Cladophora pellucida* Kütz. Low littoral and sublittoral to depths of 8m.
DOWN - Sandeel B 1972 [OM det LMI] & 1973 [OM]; Ardkeen 1975 [OM conf YMC] & 1976 [OM conf YMC]; Ballyhenry I 1975 [OM]; Kearney Pt 1982 [*Sublit Surv*]; Craiglewey 1984 (*NILS*); Orlock Pt 1985 (*NILS*); Barkley Rocks 1985 (*NILS*); Bar Hall B 1990 [OM].
ANTRIM - Rathlin I at: E Lighhouse platform 1985 (*NILS*), The Dutchman 1985 (*NILS*), Black Hd 1986 (*NILS*), Beddag 1985 (*NILS*), Arkill B 1984 (*Sublit Surv*) & Doon B 1985 (*NILS*); Ballintoy 1983 (*Sublit Surv*); Port Vinegar (*NILS*); Brown's B 1983 [*Sublit Surv*].
LONDONDERRY - Rinagree Pt 1985 (*NILS*).

Helminthora J. Ag.

Helminthora divaricata (C. Ag.) J. Ag.-**A**-
Only one old record.
ANTRIM - Antrim pre-1902 (*Batters' Cat*).

Nemalion Duby

Nemalion helminthoides (Vell.) Batt. **DAL**
Widely distributed and a common species

on the more exposed shores. Under-recorded.
DOWN - Black Neb 1985 (*NILS*); Ringburr Pt 1985 (*NILS*); Newcastle 1836 [WT]. Other unlocalised records from "Downshire" pre-1825 (JT in *Phycol Brit*) & the coast of Down pre-1841 (JT & WT in *Harvey's Man*).
ANTRIM - Portrush 1856 [WS]; White Rocks to near Portballintrae c.1933 (WSW & JWW 1933b), 1961 [MPHK]; Giant's Causeway 1853 [WS]; Colliery B 1979 [OM]; Murlough B 1898 [HH].
LONDONDERRY - Portawillin 1986 (*NILS*); Portstewart 1949 [WRM].

Scinaia Biv.

Scinaia pseudocrispa (Clem.) Cremades / **S. turgida** Chemin   **DA-**
(*Scinaia forcellata* Biv.)
S. pseudocrispa is a very rare species which, until 1926, included *S. turgida* Chemin as var. *subcostata*. The reference in *Harvey's Man* to the finding by Miss Davison of *Halymenia furcellata*, the old synonym for *S. pseudocrispa*, at Glenarm would seem to refer to *S. turgida*, for a specimen in **BEL** labelled *H. furcellata* collected by Miss Davison at Glenarm appears to be *S. turgida*. Other publications repeat this record. Indeed one (*Gifford's Mar Bot*) includes Glenarm as one of the locations and comments on the large size, distinct midrib and constrictions in the frond of the Irish specimens. These are all features which identify *S. turgida*. However those records without voucher specimens may be of either species and are listed here together.
DOWN/ANTRIM - Belfast L pre-1853 (*Gifford's Mar Bot*).
DOWN - Strangford L pre-1851 (WT in *Phycol Brit*); Portaferry pre-1902 (*Batters' Cat*).
ANTRIM - Carnlough B 1833 (Miss Davison in WT 1836); nearest coast to

The Maiden Rocks pre-1871 (GD 1871); dredged N side of Belfast L 1896 (TJ & RH 1896).

S. turgida Chemin   **DA-**
(*S. forcellata* var. *subcostata* (J. Ag.) J. Ag.)
Until 1926 this very rare species was considered to be only a variety of *S. pseudocrispa*. Records previous to 1926 may therefore be either species and have been listed in this work under: "*S. pseudocrispa/S. turgida*" qv. Only recent records and old records supported by voucher specimens are included below.
DOWN - S of Cow & Calf Dundrum B 1984 [*Sublit Surv*].
ANTRIM - Little Skerries 1982 [*Sublit Surv*]; three records from Rathlin I 1984 & 1985 [*Sublit Surv*]; Torr Hd 1983 (*Sublit Surv*); Loughan B 1983 (*Sublit Surv*); Cushendun B 1983 (*Sublit Surv*); Red B 1983 [*Sublit Surv*]; Ringfad Pt 1983 (*Sublit Surv*); Carnlough B 1983 [*Sublit Surv*]; Glenarm B pre-1841 [Miss Davison det OM]; Drumnagreagh Port 1983 [*Sublit Surv*].

Atractophora P. et H. Crouan

Atractophora hypnoides P. et H. Crouan
                                        DA-
(Incl *Rhododiscus pulcherrimus* P. et H. Crouan)
Very rare, only a few records from NI. Sublittoral to a depth of 19m.
DOWN - N of Greenore Pt 1983 (*Sublit Surv*).
ANTRIM - Rathlin I 1985 (*Sublit Surv*); Ballygalley Hd 1983 (*Sublit Surv*).

Naccaria Endlicher

Naccaria wiggii (Turn.) Endlicher   **DA-**
A very rare sublittoral species of southern shores.
DOWN - N of Greenore Pt 1983 [*Sublit Surv*].
ANTRIM - Loughan B 1983 (*Sublit Surv*).

Bonnemaisonia C. Ag.

Bonnemaisonia asparagoides (Woodw.) C. Ag.   **DAL**
The tetrasporangial phase in the life-cycle of this species is morphologically different from the gametangial phase, and indistinguishable from the tetrasporangial phase of related species. All the records below are of the gametangial phase, the tetrasporangial phase having been rarely collected in the British Is.
First recorded "Cast on shore at Donaghadee" pre-1825 [JT].
DOWN, ANTRIM and LONDONDERRY - Occasional in the lower littoral but common in the sublittoral.

B. hamifera Hariot   **DAL**
This species is an invader from Japan and was first positively determined at the end of the last century. The tetrasporangical phase is morpologically different from the gametophytic phase and known as *Trailliella intricata* Batt. It was at first considered a separate species before being recognised as a stage in the life cycle of *B. hamifera*. The tetrasporic phase of the Pacific alga - *B. nootkana* (Esper) Silva - is however also known as *T. intricata* and is indistinguishable from the tetrasporangial phase of *B. hamifera*. All records of the species from NI are of the tetrasporangial phase, so it would appear that the full sequence of stages in the life cycle does not occur here.
The first NI record is from Sandeel B 1972 (OM det LMI 1974).
DOWN, ANTRIM and LONDONDERRY - Common especially in the sublittoral.

GELIDIALES

Gelidium Lamour.

Gelidium Lamour.
Identification of the specimens of this

genus is, in some cases, difficult and the synonymy complex. PSD and LMI (*Sea B I*) recognise two aggregrate groups: *Gelidium pusillum* and *G. latifolium* and one species, *G. sesquipedale*, which is of southern distribution.

G. latifolium (Grev.) Born. et Thur.   **DAL**
Many of the specimens originally identified as *Gelidium corneum* may belong in *G. latifolium* (*Sea B I*). It is very variable in appearance but always broader than *G. pusillum*. Found in pools in the low littoral and sublittoral to depths of at least 3m. Not uncommon.
DOWN - Common.
ANTRIM - Black Hd on Rathlin I 1986 (*NILS*); Peggy's Hole 1985 (*NILS*); Giant's Causeway 1898 [HH]; Port-na-Tober 1987 (*NILS*); Portbraddan 1983 (OM); Kinbane Hd 1985 (*NILS*); Garron Pt 1976 & 1984 [OM]; near Ballintoy 1985 [OM]; Torr Hd 1985 [OM]; near Straidkilly Pt 1985 [OM]; Ballygalley Hd 1983 [*Sublit Surv*]; Seacourt 1985 (*NILS*). LONDONDERRY - Black Rock 1981 [OM].

G. pusillum (Stackh.) Le Jol.   **DAL**
(Incl *G. crinale* (Turn.) Lamour.)
Due to its variation in appearance this species is not always easy to identify. Epilithic in the mid to lower littoral.
DOWN and ANTRIM - Common.
LONDONDERRY - Portstewart 1981 [OM]; Black Rock 1981 [OM]; Rinagree Pt 1984 (*NILS*).

Pterocladia J. Ag.

[**Pterocladia capillacea** (Gmel.) Born. et Thur.   **DA-**
A southern species similar to *Gelidium*, until recently only recorded in the S and W of Ireland. Epilithic in the low littoral. All the following records are from *NILS* and are not supported by voucher specimens, they are considered doubtful

and may be of *G. latifolium*.
DOWN - Whitechurch 1986; Drummond I 1986; Carrstown Pt 1986; Mill Quarter B 1985; Craiglewey 1984; Ardglass B 1985; Ringboy 1985; Rathmullan Pt 1985; Craigalea 1980's; Bloody Br 1985; Wreck Port 1985.
ANTRIM - Rathlin I at: Portandoon 1986, Maire Isla 1986 & Portawillin 1986; Port Moon B 1986; Carrickarade 1986; Garron Harbour 1985; Halfway House 1985; Seacourt 1985.]

Gelidiella Feldm. et Hamel

Gelidiella calcicola Maggs et Guiry   **-A-**
This small creeping alga was described for the first time in 1987 (CAM & MDG 1987). There are only nine records for all Ireland, a further twelve from Britain and one from France.
Found growing on loose-lying subtidal calcareous substrata.
ANTRIM - One record only from NI: Ballygalley Hd 1983 (DWC in CAM & MDG 1987).

PALMARIALES

Palmaria Stackh.

Palmaria palmata (L.) Kuntze.   **DAL**
Dulse
P. palmata is one of the most common and best known of the red algae of our shores. It is widely distributed.
DOWN and ANTRIM - Abundant in the littoral, often epiphytic.
LONDONDERRY - Castlerock 1982 (OM); Portstewart 1854 [WS]; Black Rock sands 1915 [SW]; Rinagree B 1983 (OM) & Rinagree Pt 1984 (*NILS*).

var. **sobolifera** (Vahl) Harv.   **D—**
Among the different forms of this species is one with narrow fronds referred to by some as var. *sobolifera*. It is particularly associated with fast

flowing water.
DOWN - Ardkeen epiphytic on
Laminaria 1975 & 1976 [OM];
Ballyhenry I epiphytic on *Laminaria*
1975 [OM]; Kilclief epiphytic on
Fucus serratus 1981 [OM].

Rhodophysema Batt.

Rhodophysema elegans (P. et H. Crouan
ex J. Ag.) Dixon   **DA-**
Very rare. However as it is a small crust-
forming species it may be under-recorded.
DOWN - Sandeel B 1972 (OM det LMI);
near Kearney 1977 [OM det LMI].
ANTRIM - Antrim coast pre-1902
(*Batters' Cat*); Carnlough 1983 [OM conf
LMI].

CRYPTONEMIALES

Dilsea Stackh.

Dilsea carnosa (Schmidel) Kuntze   **DAL**
Common, but not found in quantity as is
Palmaria palmata, a species with which it
was once confused. Epilithic in the lower
littoral and upper sublittoral.
First recorded from NI in 1804 [JT].
DOWN and ANTRIM - Common.
LONDONDERRY - Near Rinagree Pt
1983 [OM].

Dudresnaya P. et H. Crouan

Dudresnaya verticillata (With.) Le Jol.
                                    DA-
Very rare with few records from NI. A
southern species of the upper sublittoral to
depths of 15m.
DOWN - Strangford L dredged 4 fathoms
1851 [WT].
ANTRIM - Torr Hd 1983 [*Sublit Surv*];
Loughan 1983 [*Sublit Surv*]; Cushendun
B 1983 [*Sublit Surv*]; Drumnagreagh Port
1983 [*Sublit Surv*].

Dumontia Lamour.

Dumontia contorta (Gmel.) Rupr.   **DAL**
(*D. incrassata* (O.F. Müll.) Lamour.)
A common epilithic species growing at
about mid-tide level.
First record from Belfast L April 1798 by
JT.
DOWN and ANTRIM - Common.
LONDONDERRY - Only three records
but probably as common in suitable sites
as in Cos Antrim & Down. Rinagree B
1983 (OM) & Rinagree Pt 1984 (*NILS*);
Balls Pt 1987 (*NILS*).

Choreocolax Reinsch

Choreocolax polysiphoniae Reinsch   **DA-**
A small parasite on *Polysiphonia lanosa*.
Probably more common than the records
suggest. The thallus is small, about the
size of a *Polysiphonia* cystocarp.
DOWN - Ballymacormick Pt 1988 [OM];
Sandeel B 1973 [OM det LMI];
Lighthouse I 1975 [OM det LMI] & 1977
[OM]; Mew I [OM conf LMI]; Rolly I
1973 (OM) & 1975 [OM det LMI];
Portavogie 1988 [OM]; Ballyhenry Pt
1975 [OM conf LMI].
ANTRIM - Torr Hd 1898 [HH]; Garron
Pt 1976 [OM conf LMI]; Ballygalley
1975 [OM conf LMI].

Harveyella Schmitz

Harveyella mirabilis (Reinsch) Schmitz et
Reinke   **D—**
This small parasite on *Rhodomela
confervoides* may well be more common
than the records indicate. Being small (the
frond is about 1-2mm across) and more or
less hemispherical in shape, it can easily
be overlooked.
DOWN - Doctor's B 1976 [OM conf
LMI] F1037; Burr Pt 1992 [OM] F10593,
F10591.

Callocolax Schmitz ex Batt.

Callocolax neglectus Schmitz ex Batt. **DA-**
A small parasite on the usually sublittoral
Callophyllis laciniata. It is not uncommon
where *C. laciniata* is found but, being
small, it is probably under-recorded.
DOWN - Sandeel B drift 1972 & 1973
[OM det LMI]; Ballyhenry I c.40ft (c.12
m) below LWM 1976 [BEP det OM conf
LMI].
ANTRIM - Murlough B 1898 [HH];
Brown's B 1904 (JA 1913); Ballygally
drift 1975 [OM det LMI]; Portmuck drift
1975 [OM det LMI], c.20ft (c.6 m) below
LWM 1976 [BEP det OM]; The Gobbins
drift 1975 [OM det LMI].

Callophyllis Kütz.

Callophyllis laciniata (Huds.) Kütz. **DA-**
A fairly common and widespread species.
Usually sublittoral and many of the older
records may be of drift material.
DOWN - Belfast L pre-1902 (*Batters'
Cat*); Groomsport [**BEL**]; Sandeel B 1972
[OM]; Donaghadee pre-1825 [JT];
Barkley Rocks 1984 (*NILS*); Springvale
1846 [WT]; Ballyhalbert 1950 [WRM];
Holm B 1984 (*NILS*); Marlfield Rocks
1985 (*NILS*); Ballyhenry I 1976 [BEP];
Audley's Castle rocks 1985 (*NILS*);
Ballyquintin 1972 [RKB]; Ballyhornan
drift 1982 (OM); Coney I 1985 (*NILS*).
ANTRIM - Common.

Kallymenia J. Ag.

Kallymenia microphylla J. Ag. **DA-**
(*Meredithia microphylla* (J. Ag.) J. Ag.)
First recorded in NI in 1982 from Garron
Pt. A southern sublittoral species found to
depths of 25m but not yet from Co
Londonderry and much more common in
Co Antrim than Co Down.
DOWN - Near Ballywalter 1982 [*Sublit
Surv*]; Lee's Wreck 1984 (*Sublit Surv*).
ANTRIM - Rathlin I : N coast 1984

(*Sublit Surv*), W Castle Hd 1984 [*Sublit
Surv*] & E Castle Hd 1985 (*Sublit Surv*),
Altachuile B 1983 [*Sublit Surv*] & 1985
(*Sublit Surv*), Altacorry Hd 1985 (*Sublit
Surv*), Farganlack Pt 1983 [*Sublit Surv*],
1984 & 1985 (*Sublit Surv*), Derginan Pt
1983 (*Sublit Surv*), Bull Pt 1984 (*Sublit
Surv*), E coast 1983 (*Sublit Surv*), Church
B 1983 & 1985 (*Sublit Surv*), White Cliffs
1984 & 1985 (*Sublit Surv*), Rue Pt 1984
& 1985 (*Sublit Surv*); Benbane Hd 1984
[*Sublit Surv*]; Larry Bane Hd 1983 [*Sublit
Surv*]; Ballintoy 1983 [*Sublit Surv*];
Kinbane Hd 1985 (*Sublit Surv*);
Ballycastle B 1985 (*Sublit Surv*) & 1989
[CAM]; Torr Hd 1983 [*Sublit Surv*] &
1984 (*Sublit Surv*); Garron Pt 1982 [BEP
& CMH] & 1983 (*Sublit Surv*); Ringfad
Pt 1983 [*Sublit Surv*]; Highlandman Rock
The Maidens 1983 [*Sublit Surv*] & 1985
(*Sublit Surv*); I of Muck 1983 [*Sublit
Surv*].

K. reniformis (Turn.) J. Ag. **DAL**
(*K. larterae* Holm.)
A very rare species of the sublittoral.
Most of the records are almost certainly of
drift material although this has not been
recorded.
DOWN - Bangor pre-1841 (*Harvey's
Man*); coast of Down pre-1846 (Miss
Davidson & WT in *Phycol Brit*);
Greencastle 1856 [WS].
ANTRIM - Portballintrae drift 1944 &
1945 (AM Irwin in HB & NFM 1953);
coast of Antrim pre-1846 (*Phycol Brit*);
Carnlough B 1833 [Miss Davison] &
"Cairnlough Bay" (=Carnlough B) 1836
(JSD 1837); nearest coast...to The Maiden
Rocks pre-1871 (GD 1871); Larne pre-
1836 (JLD in WT 1836).
LONDONDERRY - All the records
appear to be based on one old find - mouth
of the Bann pre-1836 (DM in *Fl Hib*).

Gloiosiphonia Carm.

Gloiosiphonia capillaris (Huds.) Carm.
                                        DA-
(Incl *Cruoriopsis hauchii* Batt.)
   Extremely rare with few records from all
   of Ireland. Mainly sublittoral.
   DOWN - Strangford L 1984 [CAM]
   F10641; Cloghy Rocks 1983 & 1984
   [CAM] F5289 & F10642; Newcastle Aug
   1836 [WT] F8416.
   ANTRIM - Near Glenarm pre-1841 (JLD
   in *Harvey's Man*), very probably the
   mention of this species on the "nearest
   coast ... to Maiden Rocks" (GD 1871)
   also refers to this record.

Plagiospora Kuck.

Plagiospora gracilis Kuck.   **DA-**
   An encrusting alga of the sublittoral to
   depths of 17m. No records from NI prior
   to 1983 but not easily identified and
   probably overlooked.
   DOWN - Rue Pt 1983 (*Sublit Surv*) &
   Cloghy Rocks in Strangford L narrows
   1983, 1984 & 1985 [CAM] F5290 -
   F5294.
   ANTRIM - Rue Pt Rathlin I 1985 (*Sublit
   Surv*).

Peyssonnelia Dcne.

Peyssonnelia dubyi P. et H. Crouan   **DA-**
   An encrusting thallus similar to other
   species of the genus and to some other
   genera, forming discs up to 5cm in
   diameter. Epilithic and epiphytic on rocks
   and shells but most commonly on
   encrusting corallines.
   DOWN - Foreland Pt 1986 (*NILS*); Black
   Rock near Ballywalter 1985 [OM];
   Ballywalter 1984 [OM]; Ballyhenry 1975
   [OM det LMI]; Ardkeen 1975 & 1976
   [OM conf LMI] & 1983 [OM]; Rue Pt
   1983 [CAM]; Cloghy Rocks in Strangford
   L narrows 1983 (*Sublit Surv*) & 1988
   [CAM]; Kilclief 1984 [OM]; Carlingford

L 1984 [CMH det CAM].
ANTRIM - "North of Ireland" pre-1847
(WT in *Phycol Brit* Pl. **LXXI**) - this is
included by JA in *Seaw An Co* as a Co
Antrim record; Rue Pt & Cooraghy B on
Rathlin I 1985 (*Sublit Surv*); Peggy's
Hole 1986 (*NILS*); Garron Pt 1983 (*Sublit
Surv*); Ringfad Pt 1983 (*Sublit Surv*) &
Ringfad 1984 (*NILS*); Ballygalley Hd
1983 (*Sublit Surv*); near White Hd 1988
[OM].

P. harveyana J. Ag.   **-A-**
   A very rare encrusting species, epilithic or
   epiphytic. Records are few probably
   because of the difficulty of identification.
   Sublittoral to depths of 25m.
   ANTRIM - N of Red B 1983 (*Sublit
   Surv*).

P. immersa Maggs et Irvine   **D—**
   A very rare recently described species
   forming crusts on pebbles of the low and
   sublittoral. Apparently very rare but not
   readily identified and probably under-
   recorded.
   DOWN - Cloghy Rocks in Strangford L
   narrows 1983 (*Sublit Surv*), also 1984,
   1985 & 1988 [CAM].

HILDENBRANDIALES

Hildenbrandia Nardo

Of the three species of *Hildenbrandia* in the
British Is, *H. rivularis,* is freshwater and the
other two are marine. These two species are
very similar but are distinguishable
microscopically.

Hildenbrandia crouanii J. Ag.   **DA-**
   Uncommon with few recent records.
   Littoral.
   DOWN - Rolly I 1973 [OM conf LMI];
   Doctor's B 1976 [OM conf LMI]; Black
   Is 1983 [OM det LMI] & Cloghy Rocks
   1984 [CAM] in Strangford narrows.
   ANTRIM - Colliery B 1983 [OM conf

LMI]; Carnlough 1983 [OM det LMI];
near White Hd 1988 [OM].

H. rubra (Sommerf.) Menegh.   **DAL**
More common than *H. crouanii*. Found in
the littoral and sublittoral to depths of
19m.
DOWN and ANTRIM - Common.
LONDONDERRY - Rinagree Pt 1984
(*NILS*).

CORALLINALES

Corallina L.

Corallina elongata Ellis et Sol.   **DA-**
Very rare, only one recent and confirmed
record.
DOWN - Lighthouse I 1977 [OM det
uncertain] F1210.
ANTRIM - Murlough B 1898 [HH det
uncertain OM]; Garron Pt 1976 [OM conf
LMI] F1676.

C. officinalis L.   (Pl.11) **DAL**
Very common in the mid and low littoral
extending into the upper sublittoral. In
rock pools and damp positions in
overflows from rock pools and under
Fucus. Epilithic.
DOWN and ANTRIM - Very common.
LONDONDERRY - Less common due to
lack of suitable substrate. Downhill
Strand 1984 (*NILS*); Castlerock 1982
(OM); Portstewart 1883 [DC Campbell],
1981 (OM); Black Rock 1981 (OM);
Rinagree B 1983 (OM) & Rinagree Pt
1984 (*NILS*).

Fosliella Howe

Fosliella farinosa (Lamour.) Howe   **DA-**
Very rare indeed. Only two old records of
which only one has been recently
confirmed. Possibly overlooked
DOWN - Bangor 1837 [WT].
ANTRIM - North B Portrush 1915 [SW
det YMC].

Haliptilon (Dcne.) Johansen

Haliptilon virgatum (Zanard.) Garbary et
Johansen   **D—**
(Incl *Corallina granifera* Ellis et Sol. *p.p.*)
There is only one doubtful record of two
specimens of this species from a single
site in NI.
DOWN - Bangor pre-1855 (G Johnston in
EALB 1897).

Jania Lamour.

Jania rubens (L.) Lamour.   **D—**
A very rare species. Only two old records
from NI. Epiphytic.
DOWN - Bangor October 1835 [WT];
Newcastle epiphytic on *Cladostephus
spongiosus* August 1836 [WT].

Leptophytum Adey

Leptophytum bornetii (Fosl.) Adey   **D—**
L. bornetii and *Lithothamnion foecundum*
(Kjellm.) Adey may on further research
be considered conspecific. They are
treated as separate species here and we
have no records of *L. foecundum* from NI.
There have been in the past a number of
misidentifications of this species from the
N Atlantic and Mediterranean France.
Small plants have been recorded growing
on rocks and shell debris in midlittoral to
upper sublittoral.
DOWN - Ardkeen 1983 [OM det YMC]
F4055 (YMC 1990).

Lithophyllum Phil.

Lithophyllum crouanii Fosl.   **DA-**
It has been shown that there has been a
misunderstanding concerning *L.
orbiculatum* and that the type of *L.
crouanii* Fosl. is conspecific with *L.
orbiculatum* (Fosl.) Fosl. as interpreted by
a number of workers. Thus records of *L.
crouanii* have in fact been attributed to *L.
orbiculatum* (YMC, LMI & RW 1988).

On rocks, in pools and on shells and
holdfasts of *Laminaria* just above and
below extreme low water springs.
Apparently rare but not easily identified.
DOWN - NW of Mew I 1967 (*A & A* as
L. orbiculatum (Fosl.) Fosl.); Orlock Pt
1985 (*NILS*); Foreland Pt 1986 (*NILS*);
Herring B 1985 (*NILS*); Ballyquintin Pt
1986 [*NILS* det YMC].
ANTRIM - Marie Isla on Rathlin I 1986
(*NILS*); Carrickmore 1985 [*NILS* det
YMC]; Ringfad 1984 (*NILS*);
Carnfunnock B (*NILS*).

L. fasciculatum (Lamarck) Fosl.   **-A-**
Very rare, only one record from NI which
was published under the name *L.
calcareum* (Pall.) Fosl. f. *eunana* Fosl.
ANTRIM - Larne Harbour... growing at a
depth of 7 fathoms...1898 (HH det M.
Foslie in Foslie, 1898).

L. incrustans Phil.   **DA-**
Common, but until recently under-
recorded and still no records from Co
Londonderry. Encrusting on rock and
stones in the littoral and shallow
sublittoral.
DOWN and ANTRIM - Common.

L. orbiculatum (Fosl.) Fosl.   **DA-**
As has been previously stated (under *L.
crouanii*) two distinct species have been
confused under the name *L. orbiculatum*
(YMC, LMI & RW 1988 & 1991). True
L. orbiculatum is particularly common in
rock pools (YMC - pers comm).
DOWN and ANTRIM - Common.

Lithothamnion Heydrich

Lithothamnion corallioides (P. et H.
Crouan) P. et H. Crouan   **DA-**
Very rare species with few records from
NI.
DOWN - Co Down 1898 (*Railway
Guide*).
ANTRIM - Dredged from N side of

Belfast L 4 July 1896 (TJ & RH 1896).

L. glaciale Kjellm.   (Pl.12) **DAL**
Until recently (ie the 1960's) this species
had been recorded only once in Ireland
from Larne Harbour 1902 (*BA* 1902) and
was considered as probably not
uncommon but under recorded. It is now
considered as common having been
recorded by OM, *Sublit Surv & NILS*.
On pebbles, shells, rock or surviving as
unattached fragments broken free, in the
low littoral to depths of 27m in the N
Atlantic subarctic.
DOWN and ANTRIM - Common.
LONDONDERRY - Rinagree Pt 1984
(*NILS*).

L. sonderi Hauck   **DA-**
Apparently relatively common in deep
water, but virtually unknown from the
littoral.
DOWN - Belfast L NW of Mew I 1967 (*A
& A*); Cloghy Rocks 1984 [CAM det
LMI].
ANTRIM - Larne pre-1905 (HH in Foslie,
1905).

Melobesia Lamour.

Melobesia membranacea (Esper) Lamour.
                                        DAL
Common. However being a small
encrusting epiphyte it is often overlooked.
Commonly found on *Furcellaria
lumbricalis*, *Mastocarpus stellatus*,
Laurencia pinnatifida & other algae.
DOWN and ANTRIM - Common.
LONDONDERRY - Black Rock 1917
[SW det OM]; near Rinagree Pt 1983
[OM].

Mesophyllum Lemoine

Mesophyllum lichenoides (Ellis) Lemoine
                                        DAL
A common species in the British Is
usually epiphytic on *Corallina*. Common

in the sublittoral on the Spanish coast and further south.
DOWN - Ballymacormick Pt 1984 (*NILS*); Black Neb 1985 (*NILS*); Kearney 1985 (*NILS*); Grange B 1986 (*NILS*); Mill Quarter B 1985 (*NILS*); Craiglewey 1984 (*NILS*); Ringfad Pt 1985 [OM conf YMC].
ANTRIM - Rathlin I: Church B 1983 [OM], Beddag 1985 (*NILS*) & Doon B 1985 (*NILS*); The Burnfoot 1984 (*NILS*); Peggy's Hole 1985 (*NILS*); Port Moon B [*NILS* conf YMC]; Drumnagreagh Port 1984 (*NILS*).
LONDONDERRY - Rinagree Pt 1985 (*NILS*).

Phymatolithon Fosl.

Phymatolithon calcareum (Pall.) Adey et McKibbin **DA-**
This is an important constituent of maerl or rhodolith deposits in the British Is (*A & A*). Most of the records noted below were collected from the sublittoral by diving.
DOWN - SW of Stalka Rock 1985 & N of Greenore Pt 1984 both in Carlingford L (*Sublit Surv*).
ANTRIM - Rathlin I in Church B 1985 (*Sublit Surv*); Garron Pt 1983 (*Sublit Surv*); Ballygalley Hd 1983, 1984 & 1985 (*Sublit Surv*); N side of Belfast L 1896 (TJ & RH 1896).

P. laevigatum (Fosl.) Fosl. **DA-**
Until recently, considered rare with few published records. However not readily identified and probably more common than the records suggest. Littoral and shallow sublittoral.
DOWN - NW of Mew I 1967 (*A & A*); Swineley Pt 1986 (*NILS*); Foreland Pt 1986 (*NILS*); Barkley Rocks 1984 (*NILS*); South I 1986 (*NILS*); Robin's Rock 1986 [*NILS* conf YMC]; Long Sheelah 1986 (*NILS*); Pawle I 1986 (*NILS*); The Dorn (Ardkeen) 1983 [OM det YMC], 1986 [OM det YMC] & [*NILS* conf YMC];

Ballyhenry B 1986 (*NILS*); Chapel I causeway 1986 (*NILS*); Green I 1986 (*NILS*); Granagh B 1986 [*NILS* conf YMC]; Cloghy Rocks 1984 [CAM det LMI]; Kilclief Pt 1986 (NILS); Carrstown Pt 1986 [*NILS* conf YMC]; N of Killowen 1985 (*NILS*).
ANTRIM - Rathlin I at Beddag 1985 (*NILS*) & Oweydoo 1987 (*NILS*); Peggy's Hole 1986 (*NILS*); Ringfad 1984 (*NILS*); Campbeltown 1986 (*NILS*); McIllroy's Port 1987 (*NILS*); I of Muck 1987 (*NILS*); Larne Harbour 1902 (*BA* 1902).

P. lamii (Lemoine) Chamberlain **DA-**
(*Lithophyllum lamii* Lemoine)
(*Phymatolithon rugulosum* Adey)
Actually quite common (YMC - pers comm). However not recorded from NI until recently. First recorded in the British Is in 1967.
DOWN - NW of Mew I 1967 (*A & A*); Wallaces Rocks 1984 [*NILS* det YMC]; South I 1986 [*NILS* det YMC]; Kircubbin Pt 1985 [*NILS* det YMC]; Long Sheelah 1986 [*NILS* det YMC]; Pawle I 1986 [*NILS* det YMC]; Ardkeen 1986 [OM det YMC]; Cloghy Rocks 1984 [CAM det LMI]; Craigalea 1980's [*NILS* det YMC].
ANTRIM - Black Hd on Rathlin I 1986 [*NILS* conf YMC]; The Burnfoot 1987 [*NILS* conf YMC]; Port Gorm 1985 (*NILS*); McIllroy's Port 1987 (*NILS*).

P. lenormandii (Aresch.) Adey **DAL**
A common encrusting species primarily of the littoral but rarely occurs in the sublittoral below 15m (*A & A*). All of the records from NI are from the littoral. The oldest record is from the N side of Belfast L 1896 (TJ & RH 1896).
DOWN and ANTRIM - Common.
LONDONDERRY - New Br 1987 (*NILS*); Rinagree Pt 1984 (*NILS*).

P. purpureum (P. et H. Crouan) Woelk. et
Irvine   **DAL**
(*P. polymorphum* (L.) Fosl.)
  A common encrusting alga of the more
  exposed coasts. It is a northern species
  found in the littoral and the dominant
  crustose coralline in the shallow to middle
  depths (*A &ıA*).
  DOWN and ANTRIM - Common.
  LONDONDERRY - Rinagree Pt 1984
  (*NILS*).

Pneophyllum Chamberlain

Pneophyllum concollum Chamberlain **-A-**
  Very rare, only one record from NI.
  However being small it is probable that it
  has been overlooked or confused with
  other species.
  ANTRIM - Epiphytic on *Furcellaria
  lumbricalis* in a small rock pool in the
  upper laminarian zone near the Giant's
  Causeway 1980 [OM det YMC] F2540.

P. confervicola (Kütz.) Chamberlain **DA-**
(*Fosliella minutula* (Fosl.) Ganesan)
  Very rare, although as with other small
  encrusting corallines it may be
  overlooked. Epiphytic on *Cladophora
  rupestris*, *Plocamium cartilagineum* and
  Polysiphonia elongata.
  DOWN - Rolly I drift 1975 [OM det
  YMC]; Ardkeen 1975 [OM det YMC];
  Doctor's B 1976 [OM det YMC];
  Ballyhenry I about 40 ft (c.13 m) below
  LWM 1976 [BEP det YMC]; Granagh B
  1976 [OM det YMC].
  ANTRIM - North B Portrush 1915 [SW
  det YMC]; near Giant's Causeway 1980
  [OM conf YMC].

P. fragile Kütz.   **-A-**
(*P. lejolisii* (Rosan.) Chamberlain)
(*Fosliella lejolisii* (Rosan.) Howe)
  Very rare. Only three old records of
  which only one has been confirmed.
  Epiphytic, possibly overlooked.
  ANTRIM - Drift, North B Portrush 1915

[SW det YMC]; Larne Harbour 1898
[HH]; N side of Belfast L 1896 (TJ & RH
1896).

P. microsporum (Rosenv.) Chamberlain
                                                D—
  A small encrusting alga rarely recorded in
  Ireland and only once in NI when three
  separate specimens were collected
  growing epiphytically in one large, but
  shallow, pool among boulders in the
  midlittoral.
  DOWN - Near Kearney on *Furcellaria
  lumbricalis* 1977 [OM det YMC] F1733,
  F1738a & F1948a.

Titanoderma Näg.

Titanoderma corallinae (P. et H. Crouan)
Woelk., Chamberlain et Silva   **DAL**
(*Dermatolithon corallinae* (P. et H. Crouan)
Fosl.)
  This small encrusting alga has, until
  recently, been grossly under-recorded.
  The *Prov Atlas* shows only two Irish
  records and few from Britain. Epiphytic
  on *Furcellaria lumbricalis* and probably
  also on other algae.
  DOWN - Swineley Pt 1986 (*NILS*);
  Wallaces Rocks 1984 (*NILS*); Black Neb
  1985 (*NILS*); near Kearney 1977 [OM det
  YMC] F1738c & F1948d; Down Rock
  1986 (*NILS*); Kilclief Pt 1986 (*NILS*);
  Ringfad Pt 1985 [OM det YMC] F5838a;
  St John's Pt 1985 (*NILS*); Wreck Port
  1985 (*NILS*); Greencastle Rocks 1985
  (*NILS*).
  ANTRIM - Common.
  LONDONDERRY - Rinagree Pt 1984
  (*NILS*).

T. pustulatum (Lamour.) Näg.
(*Dermatolithon pustulatum* (Lamour.) Fosl.)

  var. **pustulatum**   **DAL**
    A small encrusting epiphyte on,
    amongst other species, *Furcellaria
    lumbricalis*, *Mastocarpus stellatus* and

Laminaria holdfasts, often overgrowing *Melobesia membranacea*. Until recently there were only two records from NI and it has been greatly under-recorded. DOWN and ANTRIM - Common. LONDONDERRY - Rinagree Pt 1984 (*NILS*).

var. **confine** (P. et H. Crouan) Chamberlain   **-A-**
(*Titanoderma confinis* (P. et H. Crouan) Price, John et Lawson)
Previously confused under other names. Mainly epiphytic on various algae in the littoral, also on rocks and shells in the sublittoral (YMC 1991). ANTRIM - Epiphytic on *Mastocarpus stellatus* Portrush 1915 [SW det YMC] F455c.

var. **macrocarpum** (Rosan.) Chamberlain   **D—**
(*Tenarea hapalidioides* (P. et H. Crouan) Adey & Adey)
Only one record from NI, no doubt because it usually occurs in the sublittoral DOWN - Belfast L NW of Mew I 1967 (*A & A*).

AHNFELTIALES

Ahnfeltia Fries

Ahnfeltia plicata (Huds.) Fries   **DAL**
(Incl *Porphyrodiscus simulans* Batt.)
A frequent and distinctive species of the littoral and sublittoral to depths of 22m. The encrusting tetrasporic phase, previously referred to as *Porphyrodiscus simulans*, is apparently very rare. However the small crusts are not easily identified and may be passed unnoticed. DOWN and ANTRIM - Common. LONDONDERRY - Castlerock 1982 (OM); Black Rock 1915 [SW det OM].

GIGARTINALES

Schizymenia J. Ag.

Schizymenia dubyi (Chauv. ex Duby) J. Ag.   **DA-**
Rare, in the low littoral and sublittoral to depths of at least 17m.
ANTRIM/DOWN - Belfast B in10 fathoms pre-1852 (WT in *Phycol Brit*) and pre-1874 (*BA* 1874).
ANTRIM - Bull Pt 1984 [*Sublit Surv*]; Portrush 1800's [CA Johnson]; Cushendall 1865 [CA Johnson]; Carnlough B 1833 (Miss Davison in *Phycol Brit*); Glenarm pre-1879 (DM in *Nat Prins*), 1884 [GD]; nearest coast to Maiden Rocks c.1869 (GD); Black Hd dredged 1844 [GCH det OM]; Castle Chichester B dredged 1844 [GCH det OM].

Gracilaria Grev.

Gracilaria verrucosa (Huds.) Papenf.   **DAL**
Common and widely distributed from the littoral into the sublittoral to depths of 25m. It is tolerant of sand cover and changes in salinity.
First recorded in NI from the "Shore of the Co Antrim" pre-1825 [JT].
DOWN, ANTRIM and LONDONDERRY - Common.

Schmitzia Silva

Schmitzia hiscockiana Maggs et Guiry
(Pl.14) **DA-**
This was described as a new species in 1985 (CAM & MDG 1985) and has now been recorded from islands off the W coast of Scotland, NE, W & S Ireland, Wales, Devon and the Scilly Is. Its habitat is usually exposed to strong wave action or currents in the sublittoral at depths of 4 - 25m.
DOWN - Lighthouse I 1984 [*Sublit Surv*];

near Rue Pt 1983 [*Sublit Surv*].
ANTRIM - At several sites on Rathlin I
1984 & 1985 [*Sublit Surv*]; Murlough B
1985 (*Sublit Surv*); Torr Hd 1983 (*Sublit Surv*); Loughan B 1983 [*Sublit Surv*];
Ringfad Pt 1983 [*Sublit Surv*].

S. neapolitana (Berth.) Lagerh. ex Silva
DA-
Until recently this species was known
from the Mediterranean and Brittany but
not from the British Is. It is now known
from SW Britain and NI in the sublittoral
at depths of between 5 and 15m.
DOWN - The first Irish record is from N
of Greenore Pt in Carlingford L 1983
[*Sublit Surv*].
ANTRIM - Torr Hd 1983 [*Sublit Surv*];
Loughan B 1983 [*Sublit Surv*];
Cushendun B 1983 [*Sublit Surv*].

Gymnogongrus Martius

Gymnogongrus crenulatus (Turn.) J.
Ag. / **Gymnogongrus** sp.   **DA-**
Until recently *G. devoniensis* (Grev.)
Schotter was considered to be
synonymous with *G. crenulatus* (*Sea B I*).
However recent discoveries led to
the decision to treat them as separate
species (MDG *et al.* 1981) and later to
transfer *G. devoniensis* to a new genus
as: *Ahnfeltiopsis devoniensis* (Grev.) Silva
et De Cew (CAM - pers comm). In
addition further specimens of
Gymnogongrus have been found which
are considered a separate, and as yet
unidentified, species (CAM *et al.* 1992).
For these reasons the records below may
refer to specimens which would now be
identified as *G. crenulatus, A. devoniensis*
or the as yet unidentified species of
Gymnogongrus. However there are
probably no records of *A. devoniensis*
from NI (CAM - pers comm).
DOWN - Down pre-1977 (*Sea B I*).
ANTRIM - Antrim pre-1848 (DM in
Phycol Brit), pre-1853 (Landsborough in

Gifford's Mar Bot), pre-1874 (*BA* 1874),
pre-1902 (*Batters' Cat*); nearest coast to
The Maiden Rocks pre-1871 (GD 1871).

G. griffithsiae (Turn.) Martius   **D—**
Very rare indeed. This species has a
southern distribution in the British Is. In
Ireland it extends northwards to Co Mayo
(*Sea B I*). However there are two further
records: one from L Swilly in Co
Donegal, outside NI, collected by WS
[**BEL**]. The other is from Co Down.
DOWN - Rostrevor c.1932 (JMW 1932).

Gymnogongrus sp.   **D—**
This is the unidentified species referred to
above. It may be *G. leptophyllus* J. Ag.
(CAM *et al.* 1992).
First recorded in NI 1975.
DOWN - Ballyhenry I 1975 [OM det
CAM] F867, F868; Strangford Harbour
1986 [CAM] F5296; Rue Pt 1983 & 1986
[CAM] F5275, F5298, F5299; Cloghy
Rocks 1986 [CAM] F5276b, F5295,
F5297.

Phyllophora Grev.

Phyllophora crispa (Huds.) Dixon   **DAL**
Not uncommon in shaded positions in
rock pools of the lower littoral but
common in the sublittoral.
First recorded in NI at Bangor 1809 [JT].
DOWN and ANTRIM - Common.
LONDONDERRY - Portstewart B drift
1976 [J Burgoyne det OM]; Rinagree Pt
1984 (*NILS*).

P. pseudoceranoides (Gmel.) Newr. et A.
Tayl.   **DA-**
A common alga of pools in the lower
littoral and in the sublittoral to 17m.
Widely distributed, but no records from
Co Londonderry.
First recorded in NI in Belfast L pre-1825
[JT].
DOWN and ANTRIM - Common.

Erythrodermis Batt.

Erythrodermis traillii (Holm. ex Batt.)
Guiry et Garbary   **DAL**
(*E. allenii* Batt.)
(*Phyllophora traillii* Holm. ex Batt.)
   Uncommon and small, but not rare if
   looked for. Found only in shaded rock
   pools of the lower littoral and in the
   sublittoral to depths of 28m.
   DOWN - Common.
   ANTRIM - Rathlin I 1983, 1984, 1985
   [*Sublit Surv*] & Bruce's Cave 1987
   (*NILS*); The Skerries 1985 [*Sublit Surv*];
   Portrush 1984 (*Sublit Surv*); Peggy's Hole
   1986 (*NILS*); Port Gorm 1985 (*NILS*);
   between Gt Stookan & Gd Causeway
   1980 [OM]; Carricknaford 1987 (*NILS*);
   Larry Bane B 1984 (*NILS*); Carrickarade
   1986 (*NILS*); near Ballycastle 1984
   [*Sublit Surv*]; E of Fair Hd 1984 (*Sublit
   Surv*); Torr Hd 1898 [HH] - this was the
   first report from Ireland (HH 1899);
   Garron Pt 1976 [OM]; Ballygalley Hd
   1983 [*Sublit Surv*]; Riding Stone 1986
   (*NILS*); The Gobbins & I of Muck 1983
   [*Sublit Surv*]; Black Hd 1986 (*NILS*).
   LONDONDERRY - Rinagree Pt 1984
   (*Sublit Surv*) & (*NILS*).

Coccotylus Kütz.

Coccotylus truncatus (Pall.) Wynne et
Heine   **DAL**
(*Phyllophora truncata* (Pall.) Zinova)
   A northern species, uncommon in the
   littoral, but in the sublittoral to 16m.
   Epilithic. Reproductive outgrowths were
   at one time thought to be a separate
   parasitic species and were given the name
   Actinococcus subcutaneus (Lyngb. ex
   Hornem.) Rosenv.
   ANTRIM/DOWN - Belfast B pre-1841
   (WT in *Harvey's Man*).
   DOWN - Holywood early 1900's [MWR
   & conf ADC]; Bangor drift 1835 (WT); S
   Belfast L pre-1902 (*Batters' Cat*); Orlock
   Pt 1985 (*NILS*); Barkley Rocks 1984

(*NILS*); Strangford L pre-1836 (*Fl Hib*);
Ballyhenry B 1986 (*NILS*); Annalong
1982 [*Sublit Surv*]; Craiglewey 1984
(*NILS*); S of Nicholson's Pt 1985 (*NILS*);
Greencastle Rocks 1985 (*NILS*).
ANTRIM - Rathlin I 1983 & 1985 [*Sublit
Surv*]; N of The Storks 1984 [*Sublit Surv*];
Cushendun 1984 [*Sublit Surv*]; Garron
Harbour 1985 (*NILS*); The Maiden Rocks
pre-1908 (GD in JA 1908); Larne pre-
1836 (JLD in *Fl Hib*), 1837 [**BEL**] &
1898 [HH]; Brown's B 1983 [*Sublit
Surv*]; I of Muck 1983 [*Sublit Surv*];
Riding Stone 1986 (*NILS*); N Belfast L
dredged 1896 (TJ & RH 1896).
LONDONDERRY - Downhill 1855
[WS]; mouth of R Bann pre-1836 (DM in
Fl Hib).

Schottera Guiry et Hollenberg

Schottera nicaeensis (Lamour. ex Duby)
Guiry et Hollenberg   **DA-**
   Very rare in the littoral but not
   uncommon in the sublittoral to depths of
   27m.
   First recorded in NI at Portmuck 1975
   [OM det MDG].
   DOWN - Near Mew I & near Lighthouse
   I 1984; near Ballywalter 1982 & 1984;
   Dunnyneill Is 1983; SE of Kearney Pt
   1982; near North Rocks 1982; SE Killard
   Pt 1983; Legnaboe & Wee Pill both near
   Ardglass 1982 [all *Sublit Surv*].
   ANTRIM - At several sites on Rathlin I:
   Farganlack Pt, Altachuile B, E of Castle
   Hd, Derginan B, Bull Pt, White Cliffs &
   W Church B all between 1983 & 1985; N
   of Giant's Causeway 1982; Gid Pt 1983;
   near Larry Bane Hd 1983; SE of Kinbane
   Hd 1984; Garron Pt 1982, 1983;
   Drumnagreagh Port 1983; Ballygalley Hd
   1983 [all *Sublit Surv*]; Portmuck 1975
   [OM det MDG].

Stenogramme Harv.

Stenogramme interrupta (C. Ag.) Mont.
ex Harv.   (Pl.13) **DA-**
Very rare in the littoral, but not
uncommon in the sublittoral down to
30m. Epilithic on small stones associated
in some cases with strong currents.
DOWN - Ballywalter 1982 (*Sublit Surv*);
Pawle I 1983 (*Sublit Surv*); Dunnyneill Is
1982 (*Sublit Surv*) & 1983 [*Sublit Surv*];
Marlfield B 1982 [*Sublit Surv*] & 1983
(*Sublit Surv*); Lee's Wreck 1980 [CAM]
& 1984 (*Sublit Surv*); Ballyhenry B 1982
(*Sublit Surv*); dredged in Strangford L
near Portaferry 1858 (GD in *Nat Prins*);
Castleward B 1860 [GD]; Audley's Pt
1983 (*Sublit Surv*); Granagh B 1983
(*Sublit Surv*); Cloghy Rocks 1983 [*Sublit
Surv*].
ANTRIM - Common (*Sublit Surv*).

Mastocarpus Kütz.

Mastocarpus stellatus (Stackh.) Guiry
                                              DAL
(*Gigartina stellata* (Stackh.) Batt.)
(*Petrocelis cruenta* J. Ag.)
Carragheen, Irish Moss
Not distinguished from *Chondrus crispus*
by those who collect Carragheen. Very
common and found on most shores of Cos
Antrim and Down especially those which
are relatively exposed. However few
records from Co Londonderry. From
midlittoral in rock pools through low
littoral into upper sublittoral on exposed
rock.
Petrocelis cruenta J. Ag. is now known to
be the encrusting tetrasporic phase.
DOWN and ANTRIM - Common.
LONDONDERRY - New Br & Balls Pt
in L Foyle both 1987 (*NILS*); Downhill
Strand 1984 (*NILS*); Castlerock 1982
(OM); Portstewart 1854 [G], 1981(OM);
Black Rock 1917 [SW?]; Rinagree 1983
[OM] & 1984 (*NILS*).

Chondrus Stackh.

Chondrus crispus Stackh.   **DAL**
Carragheen or Irish Moss
A very common and widespread species
from the midlittoral into the sublittoral.
Although there are many records from
Cos Down and Antrim there are few from
Co Londonderry, no doubt due to the lack
of rocky shore. A highly variable species
according to its position on the shore and
exposure, infra-specific names have been
applied to some of these varieties. The
species is collected, along with
Mastocarpus stellatus as Carragheen or
Irish Moss, and from these "carrageenan"
is extracted for use in the stiffening of
dairy products.
DOWN and ANTRIM - Very common.
LONDONDERRY - Culmore
embankment 1937 - 39 (HB); Portstewart
1981 (OM); Black Rock 1981 (OM).

Gigartina Stackh.

Gigartina acicularis (Roth) Lamour. **D/A-**
An exceedingly rare species with only one
specimen in **BEL** from NI bearing this
name which appears to be incorrectly
determined. All the published records
appear to refer to an earlier single find
made near Belfast by JT but not recently
confirmed.
A species of southern distribution in the
British Is.
ANTRIM/DOWN - "Found ... at Belfast
by Mr Templeton, according to Mr
Turner, to whom we are obliged for a
specimen." pre-1810 (*Eng Bot*).

Polyides C. Ag.

Polyides rotundus (Huds.) Grev.   **DAL**
A common species found all around the
coast in rock pools of the littoral and into
the sublittoral to depths of 15m. Tolerant
of sand.
First recorded in NI from Belfast L pre-

1825 [JT].
DOWN, ANTRIM and
LONDONDERRY - Common.

Plocamium Lamour.

Plocamium cartilagineum (L.) Dixon
Cockscomb **DAL**
A common species on most shores in the
littoral and into the sublittoral to at least 6
- 8m. As it is attractive and not small it
has been commonly collected and is well
represented in herbaria. First recorded in
NI on the shore of the Isle of Magee 1808
[JT].
DOWN and ANTRIM - Common.
LONDONDERRY - Castlerock 1982
(OM); Portstewart B drift 1976 [J
Burgoyne]; Rinagree Pt 1984 (*NILS*).

Sphaerococcus Stackh.

Sphaerococcus coronopifolius Stackh.
DA-
(Incl *Haematocelis fissurata* P. et H.
Crouan)
Very rare, if at all, in the littoral but quite
common in the exposed sublittoral
regions of Co Antrim, especially Rathlin
I, to depths of 25m. Probably all the old
records are of drift material. There are no
recent records of littoral specimens.
First recorded in NI near Bangor 1804
[JT].
DOWN - Belfast L pre-1902 (*Batters'
Cat*); near Bangor 1804 [JT] & 1835
[WT]; Annalong 1851 [WT]. Probably all
drift specimens.
ANTRIM - Quite common in the
sublittoral [*Sublit Surv*].

Furcellaria Lamour.

Furcellaria lumbricalis (Huds.) Lamour.
DAL
Common and widespread in rock pools of
the low littoral and in the sublittoral.
Apparently less common in Co

Londonderry no doubt due to the shortage
of rocky shore.
The first record from NI is from the
Whitehouse shore 1809 [JT].
DOWN, ANTRIM and
LONDONDERRY - Common.

Halarachnion Kütz.

Halarachnion ligulatum (Woodw.) Kütz.
DA-
The foliose phase of this species varies
considerably and some of the varieties
have been named, however they are of
doubtful taxonomic validity (*Sea B I*).
The tetrasporic phase is crustose and
apparently indistinguishable from *Cruoria
rosea* (P. et H. Crouan) P. et H. Crouan
(*Sea B I*).
Very rare in the littoral but not
uncommon in the sublittoral to a depth of
5m.
DOWN - S of Lighthouse I 1984 [*Sublit
Surv*]; near Abbey Rock 1983 [*Sublit
Surv*]; Ringhaddy Sound 1985 [*Sublit
Surv*]; W of Rue Pt 1983 & 1984 [*Sublit
Surv*] & Cloghy Rocks 1983 [*Sublit Surv*]
both in Strangford L narrows; S Cow &
Calf 1984 [*Sublit Surv*]; Bloody Br 1851
[WT det Dr D]; on the S coast of
Carlingford L: W of Watson Rocks 1984
[*Sublit Surv*] & S of Vidal Rock 1984
[*Sublit Surv*].
ANTRIM - At several sites on Rathlin I
including Altachuile B, E coast, White
Cliffs & Rue Pt 1984 & 1985 [*Sublit
Surv*]; near Portrush 1982 [*Sublit Surv*];
Loughan B 1983 [*Sublit Surv*]; Garron Pt
1983 [*Sublit Surv*]; Ringfad 1983 [*Sublit
Surv*]; Drumnagreagh Port 1983 [*Sublit
Surv*]; nearest coast to The Maidens pre-
1871 (GD); The Gobbins 1985 [*Sublit
Surv*]

Catenella Grev.

Catenella caespitosa (With.) Irvine **DAL**
A common but small alga of the NE

Atlantic coast. Epilithic in the upper littoral in sheltered locations.
DOWN and ANTRIM - Common.
LONDONDERRY - Culmore 1854 [WS].

Calliblepharis Kütz.

Calliblepharis ciliata (Huds.) Kütz. **DAL**
Common in the sublittoral to depths of at least 17m but rare in rock pools of the lower littoral.
First recorded in NI from Belfast L pre-1825 [JT].
DOWN, ANTRIM and LONDONDERRY - Common in the sublittoral.

C. jubata (Good. et Woodw.) Kütz. **DA-**
A southern and western species found on shores of the low littoral and into the sublittoral to depths of at least 13m. Some of the old records may be of drift material. Rare, no recent records from the littoral.
ANTRIM/DOWN - Dredged Belfast L 1838 [GCH].
DOWN - Bangor 1835 [WT det OM]; Ballyholme B 1834 [GCH & WT det OM]; Strangford L 1834 [GCH & WT det OM]; near Dunnyneill Is 1983 [*Sublit Surv*]; Strangford L narrows 1983 [Sublit Surv]; near Greenore Pt & Green I in Carlingford L 1983 & 1984 [*Sublit Surv*]; Cranfield Pt 1985 [*Sublit Surv*].
ANTRIM - Larne 1898 [HH]; Whitehouse shore 1809 [JT].

Cystoclonium Kütz.

Cystoclonium purpureum (Huds.) Batt.
(Pl. 1) **DAL**
Common on the shore from the midlittoral into the sublittoral to depths of 13m except within Strangford L and Co Londonderry where it is less common possibly as there are fewer rocks and pools in the low littoral and more silt.
DOWN and ANTRIM - Common.

LONDONDERRY - Dredged Magilligan c.1938 (HB); Rinagree Pt 1984 (*NILS*).

Rhodophyllis Kütz.

Rhodophyllis divaricata (Stackh.) Papenf.
DAL
Very rare in the littoral, however apparently not uncommon in the sublittoral to depths of 25m.
DOWN - Belfast L pre-1902 (*Batters' Cat*); Holywood 19th C [anon det ADC]; Bangor 1835 [WT]; Mew I 1984 [*Sublit Surv*]; Craigbain 1982 [*Sublit Surv*]; Ballywalter 1836 [WT]; dredged Strangford L 1851 [WT]; Abbey Rock 1983 [*Sublit Surv*] & Holm B 1983 [*Sublit Surv*]; Portaferry 1836 [WT]; Rue Pt 1983 [*Sublit Surv*]; Legnaboe 1982 [*Sublit Surv*]; Wee Pill 1982 [*Sublit Surv*]; Annalong 1851 [WT]; E of Greenore Pt 1985 [*Sublit Surv*].
ANTRIM - Near Little Skerries & Black Rock in The Skerries 1982 [*Sublit Surv*]; several records from Rathlin I: Castle Hd 1985, Altachuile B 1985, near Altacorry Hd 1985, near Farganlack Pt 1983 & White Cliffs 1984 [all *Sublit Surv*]; Colliery B 1898 [**BEL**]; Torr Hd 1983 [*Sublit Surv*]; Loughan B 1983 [*Sublit Surv*]; Red B 1983 [*Sublit Surv*]; Ringfad Pt 1983 [*Sublit Surv*]; Drumnagreagh Port 1983 [*Sublit Surv*]; Larne 1898 [HH]; Carrickfergus pre-1849 (WMcC in *Phycol Brit*); N side of Belfast L 1896 (TJ & RH 1896); Whitehouse shore 1809 [JT det ADC].
LONDONDERRY - NW of Portstewart Pt 1984 [*Sublit Surv*].

Cruoria Fries

Cruoria pellita (Lyngb.) Fries **DA-**
An encrusting alga, probably not uncommon, but under-recorded.
Apparently most common in Co Down.
DOWN - Swineley Pt 1986 (*NILS*); Foreland Pt 1986 (*NILS*); Herring B 1985

(*NILS*); Kircubbin Pt 1985 (*NILS*);
Robin's Rock 1986 (*NILS*); Yellow Rocks
1986 (*NILS*); Ballyhenry B 1986 (*NILS*);
Kearney 1985 (*NILS*); Cloghy Rocks in
Strangford narrows 1984 & 1985 [CAM];
Sheepland Harbour 1985 (*NILS*); Ringboy
1985 (*NILS*); Wreck Port 1985 (*NILS*).
ANTRIM - Rathlin I 1983 [CAM];
Portrush 1930 (WSW & JWW 1933a);
White Rocks to near Portballintrae c.1933
(WSW & JWW 1933b); Carrickarade
1986 (*NILS*); Cushendall 1850 (WHH
1857).

Haemescharia Kjellm.

Haemescharia hennedyi (Harv.)
Vinogradova   **DA-**
(*Petrocelis hennedyi* (Harv.) Batt.)
or
Mastocarpus stellatus (Stackh. in With.)
Guiry (tetrasporic phase)
(*P. cruenta* J. Ag.)

Without tetrasporangia, which occur only
between November and March, these two
species previously referred to as: *P.
cruenta* and *P. hennedyi* "can easily be
distinguished by the presence of
secondary pit connections only in *P.
cruenta*" (CAM - pers comm). However
P. cruenta is now known to be the
tetrasporic phase of *Mastocarpus stellatus*
leaving *H. hennedyi* as the only Irish
species in the genus.
The records listed below are of specimens
which were either sterile or not
determined with confidence. They may
therefore be either *H. hennedyi* or the
tetrasporic phase of *M. stellatus*.
Apparently rare, however like some other
encrusting algae it is rarely recorded but
probably overlooked.
DOWN - Rockport 1979 [OM det LMI];
Luke's Pt 1982 [OM]; Lighthouse I 1975
[OM det LMI]; near Donaghadee 1984
[OM]; Kinnegar Rocks 1983 [OM]; near
Whitechurch 1985 [OM]; Ballyhenry I
1975 [OM].

ANTRIM - Murlough B 1977 [OM conf
LMI]; Ballygalley 1975 [OM].

H. hennedyi (Harv.) Vinogradova   **D—**
(*P. hennedyi* (Harv.) Batt.)

As stated above the tetrasporic phase of
Mastocarpus stellatus (*P. cruenta*) and *H.
hennedyi* can be distinguished by the
presence of secondary pit connections
only in *P. cruenta*. The records below are
confirmed determinations.
It is probably subtidal and has been
recorded at Strangford L narrows which is
exposed to strong currents but little wave
action (RTW & CAM 1989).
DOWN - Cloghy Rocks 1984 & 1985
[CAM] F5282, F5383 & F5284.

RHODYMENIALES

Cordylecladia J. Ag.

Cordylecladia erecta (Grev.) J. Ag.   **DAL**
This species at first appears to be rare in
the British Is, but some references note:
"of fairly frequent occurrence, especially
in Ireland" (JPC 1978). It would seem that
its main growing period is during the
winter months and this, along with its
small size, results in it being often
overlooked.
DOWN - Bangor 1835 (WT); Mew I
1984 (*Sublit Surv*); Ballywalter 1984
[*Sublit Surv*]; Pawle I 1983 [*Sublit Surv*];
Abbey Rock 1983 [*Sublit Surv*];
Dunnyneill Is 1982 & 1983 [*Sublit Surv*];
Granagh B 1983 (*Sublit Surv*); Ardglass
1982 [*Sublit Surv*].
ANTRIM - Rathlin I 1985 [*Sublit Surv*];
Portballintrae pre-1841 (DM in *Harvey's
Man* and other old publications).
LONDONDERRY - Portstewart Pt 1984
(*Sublit Surv*).

Rhodymenia Grev.

Rhodymenia delicatula P. Dang.   **DA-**
This small and apparently rare alga is

widely distributed in the British Is. It may be under-recorded partly because of its small size, about 2cm long, and being mainly sublittoral.

DOWN - Possibly our oldest specimen, det by MDG as: "probably *Rhodymenia delicatula*" was collected at Bangor in 1835 [WT]; South I 1986 (*NILS*); Pawle I & near Dunnyneill Is in Strangford L 1983 [*Sublit Surv*]; Craiglewey 1984 (*NILS*).

ANTRIM - Rathlin I 1984 [*Sublit Surv*]; E Carr Rocks The Skerries [*Sublit Surv*]; The Gobbins 1987 (*NILS*).

R. holmesii Ardiss.   **-A-**
(*R. pseudopalmata* var. *ellisiae* (Duby) Guiry)
Very rare, until 1982 unknown from NI. Epilithic in the sublittoral to 25m. All records from relatively exposed localities in Co Antrim, no confirmed records from Co Down as yet.

ANTRIM - Several sites in The Skerries 1982 & 1985 [*Sublit Surv*]; Ramore Hd 1985 [*Sublit Surv*]; Gid Pt 1983 (*Sublit Surv*); The Maidens 1985 (*Sublit Surv*).

R. pseudopalmata (Lamour.) Silva   **DAL**
A southern species very rare in the littoral but common in the sublittoral to depths of 32m. Most common on the more exposed coasts of Co Antrim.
First collected in NI by dredging in about 8 fathoms off Castle Chichester B in 1844 [GCH det WHH as *Rhodomenia palmetta*].
DOWN, ANTRIM and LONDONDERRY - Common in the sublittoral.

Champia Desveaux

Champia parvula (C. Ag.) Harv.   **-AL**
Very rare, although some references state it to be common, this may be due to confusion with other algae particularly *Chylocladia verticillata*. Further, it is

probable that one find is the basis of more than one record.

ANTRIM - "Co Antrim" pre-1825 (JT in *Harvey's Man*).

LONDONDERRY - "Derry" [DM], this may be the find which is recorded as: "Blackrocks, Portrush: Mr D. Moore" pre-1836 (*Fl Hib*) and again as "Port Rush Mr D. Moore" pre-1841 (*Harvey's Man*) - Black Rock is in Co Londonderry and Portrush is in Co Antrim.

Chylocladia Grev.

Chylocladia verticillata (Lightf.) Bliding   **DAL**
Common and widely distributed from the midlittoral into the sublittoral to 13m. The records are, with a few exceptions, from the more sheltered locations.
DOWN - Common.
ANTRIM - Antrim pre-1902 (*Batters' Cat*); Portballintrae pre-1871(GD 1871) & 1950 [AM Irwin]; Carrickfergus 1845 (WMcC in *Phycol Brit*).
LONDONDERRY - Moville Buoy [*Sublit Surv*], McKinney's Buoy, Middle Bank & The Gt Bank all in L Foyle 1983 (*Sublit Surv*).

Gastroclonium Kütz.

Gastroclonium ovatum (Huds.) Papenf.   **DAL**
Infrequent, in rock pools of the mid and low littoral into the upper sublittoral.
DOWN - Sandeel B 1972 (OM 1974); Donaghadee pre-1825 [JT], 1967 [OM]; Kearney 1985 (*NILS*); Isle O'Valla 1986 (*NILS*); Granagh B 1986 (*NILS*); Carrstown Pt 1986 (*NILS*); Craiglewey 1984 (*NILS*); St John's Pt 1967 [OM].
ANTRIM - Rathlin I at: Marie Isla 1986 (*NILS*), E Lighthouse platform 1985 (*NILS*), Beddag 1985 (*NILS*), Doon B 1985 (*NILS*) & Black Hd 1986 (*NILS*); near Wash Tub Portrush 1915 [SW]; near Dunluce Castle 1815 [JT]; between Gt

Stookan & Gd Causeway 1980 [OM]; Ringfad (*NILS*); between Glenarm & Carnlough 1836 (JSD 1837); Straidkilly Pt 1985 (*NILS*); nearest coast to The Maiden Rocks pre-1871 (GD 1871). LONDONDERRY - Blackrock Causeway undated [anon conf OM], Black Rock 1915 [SW], 1917 [almost certainly SW]; Rinagree Pt 1984 (*NILS*).

Lomentaria Lyngb.

Lomentaria articulata (Huds.) Lyngb.
DAL
A very common and readily identified species of the littoral and sublittoral to 14m depth in Cos Down and Antrim. However only two records from Co Londonderry. The oldest record from NI would appear to be that published in 1806 (*Eng Bot* **20**: 1574): "Gathered in full fructification in the summer at Larne,...by Mr Drummond".
DOWN and ANTRIM - Common.
LONDONDERRY - Near Rinagree Pt 1983 (OM) & 1984 (*NILS*).

L. clavellosa (Turn.) Gaill. (Pl.3) **DAL**
A common species of the low littoral and probably more common in the sublittoral to 25m.
First recorded from NI by JT at Glenarm in 1808.
DOWN, ANTRIM and LONDONDERRY - Common.

L. orcadensis (Harv.) Coll. ex W. Tayl.
DAL
A species common to below 20m in the sublittoral from which there are over 100 recent records for NI. However very rare in the littoral where it was first recorded in NI at Garron Pt in 1976 (OM 1978).
DOWN and ANTRIM - Common especially in the sublittoral.
LONDONDERRY - Three records: two from NW Portstewart Pt 1984 [*Sublit Surv*] & one from NW Portstewart Pt/

Rinagree Pt 1984 (*Sublit Surv*).

CERAMIALES

Antithamnion Näg.

Antithamnion cruciatum (C. Ag.) Näg.
DA-
Rare with few recent records. However as a small and difficult to determine species it has probably been overlooked or misidentified.
DOWN - It is recorded from the "coast of Down" in several old publications probably all based on one find pre-1841 (WT in *Harvey's Man*); Barkley Rocks 1984 (*NILS*); St John's Pt 1984 (*Sublit Surv*); Watson Rocks 1984 [*Sublit Surv*].
ANTRIM - Rathlin I 1983 (*Sublit Surv*); Garron Pt 1983 (*Sublit Surv*); Ballygalley Hd 1983 (*Sublit Surv*); nearest coast ... to The Maiden Rocks pre-1869 (GD 1871).

Antithamnionella Lyle

Antithamnionella spirographidis (Schiffner) Wollas. **DA-**
Very rare in the littoral, however several records from the sublittoral to depths of 23m. Mostly in the more exposed locations.
DOWN - Near Ballyhalbert 1982 [*Sublit Surv*]; Annalong 1982 [*Sublit Surv*].
ANTRIM - Two records from The Skerries 1982 [*Sublit Surv*]; Ringfad Pt 1983 [*Sublit Surv*]; Carnlough B 1983 [*Sublit Surv*]; Straidkilly 1985 (*NILS*).

Aglaothamnion
Feldmann-Mazoyer

The genus *Aglaothamnion* was first erected in 1941 for species of *Callithamnion* with uninucleate cells. This was not accepted by many phycologists and it is not accepted in South & Tittley (*N A Checklist*). However Maggs & Hommersand (*Sea B I*) consider that the uninucleate and multinucleate species should be placed in separate genera and therefore recognise *Aglaothamnion* as a genus distinct from *Callithamnion*. These two genera are difficult, the specimens are small and not easily determined. Many of the records, unless recently confirmed, are considered doubtful.

Aglaothamnion byssoides (Harv.)
L'Hardy-Halos et Rueness   **DAL**
(*Callithamnion byssoides* Arnott ex Harv.)
  Epilithic and epiphytic in pools and sheltered areas of the littoral into the sublittoral. Widespread.
  DOWN - Craigavad 1840 [WT]; Rolly I 1975 [OM det JP]; Audley's Castle rocks 1985 (*NILS*); Portaferry (dredged) 1838 [WT]; Isle O'Valla 1986 (*NILS*); Coney I 1985 (*NILS*); Dundrum 1836 & 1851 [WT].
  ANTRIM - Gt Stookan 1986 (*NILS*); Marchburn Port 1984 (*NILS*).
  LONDONDERRY - New Br 1987 (*NILS*).

A. hookeri (Dillw.) Maggs et Hommersand   **DAL**
(*Callithamnion hookeri* (Dillw.) S.F. Gray)
  A very variable species with a wide variation in size and branching. Found in rock pools of the low littoral growing usually epiphytically or frequently epizoically.
  DOWN and ANTRIM - Common.
  LONDONDERRY - Culmore 1854 [WS?]; mouth of the Bann 1856 [G] & [WS]; Rinagree Pt 1984 (*NILS*).

A. roseum (Roth) Maggs et L'Hardy-Halos   **D—**
(*Callithamnion roseum* (Roth) Lyngb.)
  Epiphytic and epizoic in the low littoral and sublittoral. Widespread. All recent records due to *NILS,* previous to which it was considered rare in NI possibly because it is similar to, and confused with, other species.
  DOWN - Swineley Pt 1986 (*NILS*);

Ballywalter 1836 [WT]; Kearney 1985 (*NILS*); Audley's Castle rocks 1985 (*NILS*); Ballyquintin Pt 1986 (*NILS*); Craiglewey 1984 (*NILS*); Ardglass 1851 [WT].

A. sepositum (Gunn.) Maggs et Hommensand   **DAL**
(*Callithamnion sepositum* (Gunn.) Dixon et Price)
  More common in Co Antrim than Co Down and fairly widespread on the eastern shores of the N Atlantic. Commonly epiphytic or epizoic in the littoral.
  DOWN - Ballywalter 1836 [WT]; Mahee I 1985 (*NILS*); Audley's Castle rocks 1985 (*NILS*); Ardglass 1851 [WT]; St John's Pt 1985 (*NILS*).
  ANTRIM - Common.
  LONDONDERRY - Downhill Strand 1984 (*NILS*); Rinagree Pt 1984 (*NILS*).

Callithamnion Lyngb.

Callithamnion corymbosum (Sm.) Lyngb.
                                         DA-
  Usually epiphytic on other algae and *Zostera* in the lower littoral and sublittoral in sheltered locations. Widespread but variable in growth form.
  DOWN - Cultra [**BEL**]; Black Neb inlet 1986 (*NILS*); Holm B 1984 (*NILS*); Craigalea 1980's (*NILS*); Newcastle 1836 [WT]; Annalong 1851 [WT].
  ANTRIM - Larry Bane B 1984 (*NILS*).

C. granulatum (Ducluz.) C. Ag.   **DA-**
(Incl *C. spongiosum* Harv. in Hook.)
  Rock pools of the lower littoral in relatively exposed locations. Epiphytic and epizoic. Rare.
  DOWN - Bangor 1835 & 1836 [WT]; Ardglass 1851 [WT]; Bloody Br 1851 [WT].
  ANTRIM - Rathlin I at: Black Hd & Marie Isla 1986, E Lighthouse platform 1985; The Burnfoot 1984; Campbeltown

& Whitebay Pt 1986 (all *NILS*).

C. tetragonum (With.) S.F. Gray   **DAL**
Relatively common in the littoral and
sublittoral on the more exposed shores.
Epiphytic with wide distribution.
DOWN - Bangor 1835 [WT];
Ballymacormick Pt 1984 (*NILS*);
Groomsport 1835 [WT]; Mahee I 1985
(*NILS*); Rue Pt & Cloghy Rocks in
Strangford L narrows 1983 (*Sublit Surv*);
Ardglass 1851 [WT].
ANTRIM - Rathlin I at: Portandoon 1986
(*NILS*), Marie Isla 1986 (*NILS*), E
Lighthouse platform 1985 (*NILS*) & Doon
B 1985 (*NILS*); The Skerries 1982 [*Sublit
Surv*]; White Rocks 1979 [OM]; Peggy's
Hole 1985 & 1986 (*NILS*); Port Gorm
1975 (*NILS*); The Ladle 1985 (*NILS*); Port
Moon 1984 (*NILS*); Larry Bane B 1984
(*NILS*); near Ballycastle 1978 [OM];
Murlough B 1898 [EMH]; Torr Hd 1983
(*Sublit Surv*); Garron Pt 1976 [OM conf
JP]; Portmuck 1975 (OM det JP).
LONDONDERRY - "Derry" pre-1981
(PSD & JP 1981).

C. tetricum (Dillw.) S.F. Gray   **DA-**
Epilithic or epiphytic in the littoral
especially on the more exposed shores.
First recorded in NI by JT in 1811.
DOWN - Near Bangor 1811 [JT]; Black
Neb 1985 (*NILS*); Bar Hall B 1986
(*NILS*).
ANTRIM - The Burnfoot 1984 (*NILS*);
Larry Bane B 1984 (*NILS*); Campbeltown
1986 (*NILS*); Black Hd 1986 (*NILS*).

Ceramium Roth

The species of *Ceramium* which occur on
the shores of the British Is are still
imperfectly understood. The four "spiny"
species were clearly sorted out in 1960
(PSD 1960). CAM & MHH (*Sea B I*) have
considerably advanced our understanding of
the genus but have been able to apply only
"provisional names" in some cases. Until

further information becomes available some
specimens, which require redetermination,
and published records, have been placed in
C. rubrum agg.

Ceramium ciliatum (Ellis) Ducluz.   **DA-**
This is one of the four "spiny" species of
Ceramium in the British Is and can be
identified with relative ease. Epilithic in
deep shaded rock pools of the low littoral.
DOWN - Orlock Pt 1985 (*NILS*); Black
Neb 1985 (*NILS*); Ballywaddan 1988
[OM]; Chapel I causeway (*NILS*);
Craiglewey 1984 (*NILS*); Sheepland
Harbour 1985 (*NILS*); Ardglass 1851
[WT] & 1984 [OM].
ANTRIM - Rathlin I at Marie Isla 1986
(*NILS*); Portrush 1854 [G conf OM];
White Pk B 1915 [MWR conf OM];
Murlough B 1898 [HH conf OM];
Straidkilly Pt 1985 (*NILS*).

C. cimbricum Petersen.   **D—**
(*C. fastigiatum* Harv.)
Rare. Only recorded in print from Co
Cork in Ireland (MDG 1978).
DOWN - Recorded in Strangford L
(CAM - pers comm).

[**C. derbesii** Solier ex Kütz.   **-A-**
Unlike the "spiny" species *C. derbesii* is
not readily identified and there are no
recent records of it from NI. CAM &
MHH (*Sea B I*) have not included it and
consider most of the records of it from the
British Is to be of *C. nodulosum* or *C.
ciliatum*.
ANTRIM - The only records from NI are
from Co Antrim and are based on one, or
perhaps two, finds: Murlough B 1898
(HH), published as a new record to the
British flora by Hanna (1899), the record
has been repeated in other publications
such as *BA* 1902. There are however also
two specimens in **BEL** [F6762 & F6763]
collected at the nearby Torr Hd on the 4th
August 1898 by Hanna. In his paper
Hanna comments that: "The first week in

August was spent at Murlough Bay and neighbouring localities were examined such as Colliery Bay and Torr Head."]

C. deslongchampii Chauv.   **DAL**
Although this species was included in the early checklists Parke (1953), and Parke & Dixon (1968), it was omitted from *Br Check-list* (1976) but included in *N A Checklist* (1986). As a result of the general confusion in the genus as a whole and in this entity in particular, records may have been lost or included under other names.
DOWN/ANTRIM - Belfast B pre-1825 (*Harvey's Man*).
DOWN - Common in Helen's B (CAM - pers comm); Orlock Pt 1985 (*NILS*).
LONDONDERRY - Culmore in L Foyle 1854 [WS].

C. diaphanum (Lightf.) Roth   **DA-**
(Incl *C. tenuissimum* (Roth) J. Ag.)
Some authorities consider *C. diaphanum* to be part of a *Ceramium* complex including, among others, *C. strictum* Harv. (*N A Checklist*). However CAM & MHH (*Sea B I*) record "*Ceramium strictum* sensu Harv." as a separate species but note that: "A valid name has not yet been determined for this species." They also transfer *C. tenuissimum* to *C. diaphanum.* In the records noted below, *C. strictum* may be included. However as it is a sourthern species (*Sea B I*) this is unlikely, and *C. tenuissimum* is now included in agreement with *Sea B I* .
Due to the taxonomic confusion a number of the records below require confirmation. Epilithic and epiphytic mainly in the littoral.
DOWN - Bangor 1835 [WT]; Ballyholme B dredged in 4 fathoms 1846 [GCH]; The Dorn 1986 (*NILS*); Ringburr Pt 1985 (*NILS*); near Killyleagh 1984 [OM]; Chapel I causeway 1986 (*NILS*); Portaferry dredged 1838 [WT]; Granagh B 1972 [CRT det LMI as "*C. diaphanum*

complex"]; Ardglass B 1985 (*NILS*); Rathmullan Pt 1985 (*NILS*); Newcastle 1836 [WT]; Annalong 1851 [WT] & 1983 [OM].
ANTRIM - Port Gorm 1985 (*NILS*); Slidderycove Pt 1980 [OM]; Portrush 1854 [**BEL**]; Dalaradia Pt 1986 (*NILS*); N side of Belfast L 1896 (TJ & RH 1896); Macedon Pt 1915 [SW].

C. echionotum J. Ag.   **DA-**
This is one of the four relatively easily identified "spiny" species of *Ceramium* recorded in Ireland. Not uncommon, however probably under-recorded.
DOWN - Whitechurch 1986 (*NILS*); Herring B 1985 (*NILS*); Kircubbin Pt 1985 (*NILS*); Yellow Rocks 1986 (*NILS*); Limestone Rock 1986 (*NILS*); Kearney 1985 (*NILS*); Marlfield Rocks 1985 (*NILS*); Isle O'Valla 1986 (*NILS*); Carrstown Pt 1986 (*NILS*); Ballyhornan 1982 [OM]; Ardglass 1851 [WT]; Rathmullan Pt 1985 (*NILS*); Craigalea 1980's (*NILS*); Bloody Br 1985 (*NILS*); Wreck Port 1985 (*NILS*).
ANTRIM - Portrush 1854 [**BEL**]; Port-na-Tober Hd 1987 (*NILS*); Port Moon B 1986 (*NILS*); Dunimeny Castle 1987 (*NILS*); Seacourt 1985 (*NILS*); Larne 1902 (*Batters' Cat*); Chapman's Rock 1987 (*NILS*); Barney's Pt 1986 (*NILS*); Loughshore Pk 1986 (*NILS*).

C. gaditanum (Clem.) Cremades   **DA-**
(*C. flabelligerum* J. Ag.)
One of the four spiny *Ceramium* sp. of the British Is which is readily identified under the microscope. More frequently recorded from Co Antrim than Co Down.
DOWN - "Downshire coast" pre-1853 (*Gifford's Mar Bot*); Lighthouse I 1977 & 1978 [OM]; Ballywalter 1988 [OM]; Wallaces Rocks 1984 (*NILS*); Taggart I 1986 (*NILS*); Ballyhornan 1982 [OM]; Ringfad 1985 [OM]; near St John's Pt 1988 [OM].
ANTRIM - Rathlin I at Portandoon 1986

(*NILS*); North B Portrush 1915 [SW]; Collier B 1978 & 1983 [OM]; Port Gorm 1985 (*NILS*); Torr Hd 1898 [HH]; Garron Pt 1976 & 1984 [OM]; Carrickmore 1985 (*NILS*); Cushendun 1985 (*NILS*); Red Arch 1985 (*NILS*); Garron Harbour 1985 (*NILS*); Carnlough 1983 [OM]; Straidkilly Pt 1985 [OM] & (*NILS*); Glenarm B 1983 [OM conf PSD]; Ballygalley 1975 [OM det PSD]; Seacourt 1985 (*NILS*); Chapman's Rock 1987 (*NILS*); Marchburn Port 1984 (*NILS*).

C. rubrum agg.   **DAL**
This is an illegitimate name in a genus with major nomenclatural problems (*Sea B I*). South & Tittley (*N A Checklist*) list ten species of *Ceramium* without spines from Ireland (including "*C. ?circinnatum* (Kütz.) J. Ag.") and a further four with spines. Within *C. rubrum* (Huds.) C. Ag. they include a further fifteen N Atlantic species in symonymy. CAM & MHH point out that there appear to be four British Is representatives of the "*C. rubrum* group" to which they have assigned four provisional names: "*C. pallidum, C. botryocarpum, C. nodulosum* (a valid name for the species known as *C. rubrum*), and *C. secundatum,* which has frequently been referred to '*C. pedicellatum* Duby'"(*Sea B I*). As these are provisional names and most of the specimens noted here may require redetermination, I have listed them all under "*C. rubrum* agg." until the situation is clarified and more information is available.
Most of the records consist probably of *C. nodulosum* and *C. pallidum* (-pers comm CAM).
DOWN - Common, the records include: Bangor drift 1915 [SW]; Sandeel B 1972 [OM det PSD as: *C. pedicellatum*] F3574 & [OM det PSD as *C. rubrum*] F3575; Kinnegar Rocks 1983 [OM det PSD as: *C. rubrum "rubrum"*] F3671; Ardglass 1871 [WT]; Newcastle B 1851 [WT]; Dundrum

B 1851 [WT]; Annalong 1851 [WT].
ANTRIM - Portballintrae 1949 [WRM]; near Giant's Causeway 1980 [OM det PSD as: *C. pedicellatum*] F2547; Murlough B 1915 [MWR]; Cushendun 1871 [SAB]; Carnlough 1983 [OM det PSD as: *C. rubrum "rubriforme"*] F3640; Portmuck 1975 [OM det PSD as: *C. pedicellatum*] F939; The Gobbins 1975 [OM det PSD as: *C. pedicellatum*] F899; Macedon Pt 1915 [SW].
LONDONDERRY - Balls Pt 1987 (*NILS*); Castlerock 1982 [OM]; Portstewart 1981 [OM]; Rinagree Pt 1984 (*NILS*).

C. shuttleworthianum (Kütz.) Rabenh.
                                      DAL
One of the relatively common species readily identified under the microscope by the "spines" on the cortical bands. Common on *Mytilus* in the lower littoral of exposed shores.
DOWN and ANTRIM - Common.
LONDONDERRY - Downhill Strand 1984 (*NILS*); Castlerock 1982 (OM); Black Rock 1915 [SW]; Rinagree B 1983 [OM] & Rinagree Pt 1984 (*NILS*).

Compsothamnion Näg.

Compsothamnion gracillimum de Toni
                                      DAL
Very rare and only recorded in NI from the sublittoral to depths of 15m.
DOWN - Ballywalter 1984 (*Sublit Surv*); Ringhaddy Sound 1985, Pawle I 1983 & near Black Rock in Strangford L 1983 [all *Sublit Surv*]; Lee's Wreck 1984 (*Sublit Surv*); near Ardglass 1982 [*Sublit Surv*]; near Cow & Calf 1984 & St John's Pt 1984 & 1985 [*Sublit Surv*].
ANTRIM - Rathlin I 1984 & 1985 (*Sublit Surv*).
LONDONDERRY - S of Middle Bank 1983 [*Sublit Surv*].

C. thuyoides (Sm.) Schmitz   **D—**
A very rare species in NI, however it is
probably more common in the sublittoral.
DOWN - Portaferry dredged 1838 [WT];
Strangford L pre-1874 (*BA* 1874) it is
possible that this refers to the Portaferry
record above.

Griffithsia J. Ag.

Griffithsia corallinoides (L.) Batt.   **DAL**
A rare species. Most recent records
sublittoral to depths of 20m.
DOWN - Ringhaddy Sound 1985 (*Sublit
Surv*); near Black Rock in Strangford L
1983 [*Sublit Surv*]; Holm B 1983 [*Sublit
Surv*], 1984 (*NILS*); Portaferry dredged
1838 [WT]; Audley's Pt 1983 [*Sublit
Surv*].
ANTRIM - Rathlin I 1985 (*Sublit Surv*);
Torr Hd 1983 (*Sublit Surv*); Cushendun
1984 (*Sublit Surv*); Ballygalley Hd 1983
(*Sublit Surv*).
ANTRIM/DOWN - Belfast L dredged
1839 [EG].
LONDONDERRY - Derry pre-1879
[DM] & Black Rock pre-1836 (DM in *Fl
Hib*) - these may refer to the same find.

Halurus Kütz.

Halurus equisetifolius (Lightf.) Kütz. **-AL**
Sea Tail
Rarely recorded in NI. Apparently more
common on the exposed shores of Co
Antrim. Littoral and sublittoral to depths
below 3m.
ANTRIM - Rathlin I at Doon B 1985
(*NILS*); Portrush 1853 [WS] F3344 &
1902 (JA 1904b); Slidderycove Pt 1980
[OM] F2447; The Burnfoot 1984 (*NILS*);
Giant's Causeway 1853 [WS] F1528;
Ballintoy 1983 [*Sublit Surv*] F4552 &
F4554.
LONDONDERRY - "Derry" pre-1852
[Moon] F8460.

H. flosculosus (Ellis) Maggs et
Hommersand   **DAL**
(*Griffithsia flosculosa* (Ellis) Batt.)
A common species. Possibly because it is
readily determined, there are more records
of it than some other, equally common but
less easily determined, species.
DOWN and ANTRIM - Common.
LONDONDERRY - Rinagree Pt 1984
(*NILS*).

Monosporus Solier

Monosporus pedicellatus (Sm.) Solier
                                             DA-
(*Corynospora pedicellata* (Sm.) J. Ag.)
A very rare sublittoral species for which
there are few recent records from NI.
DOWN - Bangor 1835 [WT]; Belfast B
(WT in *Harvey's Man*); Audley's Pt 1983
[*Sublit Surv*]; near Dunnyneill Is 1983
[*Sublit Surv*]; Portaferry pre-1841 (WT in
Harvey's Man).
ANTRIM - Belfast pre-1902 (*Batters'
Cat*).

Plumaria Schmitz

Plumaria plumosa (Huds.) Kuntze   **DAL**
(*Plumaria elegans* (Bonnem.) Schmitz)
Common. However only one record from
Co Londonderry.
DOWN and ANTRIM - Common.
LONDONDERRY - Near Rinagree Pt
1983 (OM); Rinagree Pt 1984 (*NILS*).

Pterothamnion Näg.

Pterothamnion plumula (Ellis) Näg. **DAL**
(*Antithamnion plumula* (Ellis) Thur.)
Relatively rare in the littoral but common
in the sublittoral. First recorded in NI at
Holywood in 1835 where it was: "Thrown
ashore in great abundance..." (WT 1836).
Records of it occur in 1836, 1838, 1845,
1851 in other localities in NI. All the
recent records are from the sublittoral to
depths of 17m.

DOWN, ANTRIM and LONDONDERRY - Common in the sublittoral.

Ptilota C. Ag.

Ptilota gunneri Silva, Maggs et Irvine **DAL**
(*Ptilota plumosa* (Huds.) C. Ag.)
A common northern species, however with few records from Co Londonderry.
DOWN and ANTRIM - Common.
LONDONDERRY - Magilligan 1833 [GCH]; Portstewart 1883 [DC Campbell], 1976 [J Burgoyne]; Black Sands (drift) 1915 [SW].

Ptilothamnion Thur.

Ptilothamnion pluma (Dillw.) Thur.   **-A-**
Very small, less than 2cm in length and very rare. Reported in the literature from *Laminaria* stipes.
ANTRIM - Rathlin I at Portandoon 1986 (*NILS*).

Seirospora Harv.

Seirospora interrupta (Sm.) Schmitz **D—**
(*Seirospora seirosperma* (Harv.) Dixon)
Very rare. Only one recent record from NI and few records for all Ireland.
DOWN - Holm B 1983 (*Sublit Surv*); Portaferry dredged July 1838 [WT].

Spermothamnion Aresch.

[Spermothamnion mesocarpum (Carm. ex Harv.) Chemin   **-A-**
CAM & MHH (*Sea B I*) have been unable to confirm the validity of this species and have excluded it.
Extremely rare. Only one record from NI.
ANTRIM - N side of Belfast L dredged 1896 (TJ & RH 1896 as *Rhodochorton mesocarpum*).]

S. repens (Dillw.) Rosenv.   **DA-**
Rare with no recent records from the

littoral. However there are recent records from the sublittoral to depths of 24m.
DOWN - Bangor 1837 & 1846 [WT]; Abbey Rock 1983 [*Sublit Surv*]; Ardglass 1851 [WT]; St John's Pt 1984 [*Sublit Surv*]; Newcastle 1836, 1838 & 1851 [WT]; Annalong 1851 [WT] & 1982 (*Sublit Surv*).
ANTRIM - N of Red B 1983 (*Sublit Surv*); Barney's Pt 1986 (*NILS*).

Sphondylothamnion Näg.

Sphondylothamnion multifidum (Huds.) Näg.   **DA-**
Very rare with few recent records all of which are sublittoral.
DOWN/ANTRIM - Belfast L (WT in *Phycol Brit*); other published records are based on this, eg *Seaw An Co*, in which the record is considered to be from Co Antrim.
DOWN - Bangor 1835 [prob coll WT]; near Roe I 1983 [*Sublit Surv*]; near Black Rock in Strangford L 1983 [*Sublit Surv*].
ANTRIM - Rathlin I: Rue Pt 1984 [*Sublit Surv*], Altachuile B 1985 (*Sublit Surv*), E coast 1984 [*Sublit Surv*] & 1985 (*Sublit Surv*).

Spyridia Harv.

Spyridia filamentosa (Wulf.) Harv.   **D—**
A very rare species of southern distribution in the British Is but possibly more common than this single record indicates.
DOWN - E of Black Rock in Strangford L sublittoral 1983 [*Sublit Surv*] F4915.

Acrosorium Zanard.

Acrosorium venulosum (Zanard.) Kylin
                                         -A-
(*A. uncinatum* Kylin)
Extremely rare. The difficulty of distinguishing *A. venulosum* from *Cryptopleura ramosa* is such that records

may have been lost altogether. Possiby *A. venulosum* is more widespread in the low and sublittoral.
ANTRIM - Port Gorm 1985 (*NILS*); Murlough B 4 August 1898 [**BEL**] probably the records quoted in *BA* 1902 & *Batters' Cat* are based on this specimen or its duplicates; Garron Harbour 1985 (*NILS*); Halfway House 1985 (*NILS*).

Apoglossum J. Ag.

Apoglossum ruscifolium (Turn.) J. Ag.
DA-
Rare in the littoral but not uncommon in the sublittoral to depths of 25m.
DOWN - Near Mew I 1984 [*Sublit Surv*]; Strangford L (dredged) 4 fathoms 1851 [WT] & Strangford L narrows 1983 [*Sublit Surv*]; Killough B 1982 [*Sublit Surv*]; Annalong 1851 [WT]; Kilkeel 1982 [*Sublit Surv*].
ANTRIM - The Skerries 1982 [*Sublit Surv*]; several sites on Rathlin I: Altachuile B 1983 [*Sublit Surv*], Farganlack Pt 1985 [*Sublit Surv*] & W Lighthouse 1983, Arkill B 1984, Bull Pt 1984, E coast 1984, W Rue Pt 1984 & 1985, Castle Hd 1984, Cooraghy B 1985, Altacorry Hd 1985 & Derginan Pt 1985 (all *Sublit Surv*); Loughan B 1983 [*Sublit Surv*]; Ringfad 1983 [*Sublit Surv*].

Cryptopleura Kütz.

Cryptopleura ramosa (Huds.) Kylin ex Newton
DAL
(Incl *Acrosorium reptans* (P. et H. Crouan) Kylin)
An abundant species especially of the lower littoral in rock pools, among *Laminaria* holdfasts and in the sublittoral.
DOWN and ANTRIM - Abundant.
LONDONDERRY - Very few records, but probably abundant. Castlerock drift 1982 (OM); Black Rock 1915 [SW]; near Rinagree Pt 1983 [OM].

Delesseria Lamour.

Delesseria sanguinea (Huds.) Lamour.
(Pl.15) **DAL**
A common species of shaded positions in rock pools of the low littoral and sublittoral.
DOWN and ANTRIM - Common.
LONDONDERRY - Magilligan c.1951 (HB).

Drachiellla Ernst et Feldm.

Drachiella heterocarpa (Chauv. ex Duby) Maggs et Hommersand   **-A-**
(*Myriogramme heterocarpum* (Chauv. ex Duby) Ernst et Feldm.)
Unknown in the littoral, however recorded from the sublittoral to depths of 28m in the *Sublit Surv*. The records below result from this survey and except for Cooraghy B are supported by specimens in **BEL**.
ANTRIM - The Skerries 1982; Cooraghy B, Derginan Pt & Doon Pt on Rathlin I 1985; Torr Hd 1983 [*Sublit Surv*].

D. spectabilis Ernst et Feldm.   **-A-**
A rare and inconspicuous sublittoral species. The only NI records are from the *Sublit Surv*.
ANTRIM - Rathlin I at: W of Church B 1983, Farganlack Pt 1984, E of Castle Hd 1985 & Black Hd 1985; near Kinbane Hd 1984 & 1985 [all *Sublit Surv*].

Gonimophyllum Batt.

Gonimophyllum buffhamii Batt.   **-A-**
Extremely rare. A small parasite on *Cryptopleura ramosa*.
ANTRIM - Garron Pt 1976 [OM] F1629; Portmuck 1976 [BEP det OM] F802; N side of Belfast L 1896 (TJ & RH 1896).

Hypoglossum Kütz.

Hypoglossum hypoglossoides (Stackh.)
Coll. et Harv. **DAL**
(*H. woodwardii* Kütz.)
Uncommon but, like others of the family,
an attractive plant and often preserved.
DOWN/ANTRIM - Near Belfast 1830
(JLD in *Alg Brit*).
DOWN - South I 1986 (*NILS*); Rolly I
drift 1975 [OM]; Doctor's B 1976 [OM];
Ardkeen 1983 [OM]; N of Killowen 1985
(*NILS*).
ANTRIM - Rathlin I E of Lighthouse
platform 1985 (*NILS*); Doon B 1985
(*NILS*); Killeaney 1980's (*NILS*);
Slidderycove Pt 1980 [OM];
Portballintrae 1949 [WRM]; Murlough B
1915 [MWR]; Red B dredged 1842
[GCH]; Cairnlough (=Carnlough) B 1836
(JSD) & 1842 [WT]; nearest coast to The
Maiden Rocks pre-1871 (GD); near Larne
1804 [JT], Larne 1835 [JLD], 1898 [HH]
& Larne L 1841 [GCH].
LONDONDERRY - Mouth of L Foyle
c.1850 [WS]; Rinagree Pt 1984 (*NILS*).

Membranoptera Stackh.

Membranoptera alata (Huds.) Stackh.
DAL
Common in the low littoral and in the
sublittoral. Epilithic and epiphytic.
DOWN and ANTRIM - Common.
LONDONDERRY - Portstewart B 1976
[J Burgoyne]; Black Rock 1915 [SW];
Rinagree B 1983 (OM) & Rinagree Pt
1984 (*NILS*).

Haraldiophyllum Zinova

Haraldiophyllum bonnemaisonii (Kylin)
Zinova **DAL**
(*Myriogramme bonnemaisonii* Kylin)
Rare in the littoral. However not
uncommon especially, it seems, in Co
Antrim in the sublittoral to depths of 27m.
DOWN - Mew I 1984 [*Sublit Surv*]; near

Ballywalter 1982 [*Sublit Surv*] & 1984
(*Sublit Surv*); Audley's Pt 1983 (*Sublit
Surv*) & Granagh B 1983 (*Sublit Surv*);
Strangford L narrows 1983 [*Sublit Surv*];
Portaferry 1835 [GCH & WT]; Killough
B 1982 [*Sublit Surv*].
ANTRIM - Rathlin I 1983 & 1984 [both
Sublit Surv]; Portrush 1985 [*Sublit Surv*];
The Skerries 1982 [*Sublit Surv*], 1984 &
1985 (both *Sublit Surv*); near Giant's
Causeway 1982 [*Sublit Surv*]; Benbane
Hd 1984 (*Sublit Surv*); Gid Pt 1983
(*Sublit Surv*); Larry Bane Hd 1983 [*Sublit
Surv*]; Kinbane Hd 1984 (*Sublit Surv*);
Murlough B 1898 [HH] & 1915 [MWR];
Torr Hd 1983 [*Sublit Surv*]; Ringfad Pt
1983 (*Sublit Surv*); nearest coast to The
Maidens pre-1871 (GD 1871) & 1985
(*Sublit Surv*); Larne 1824, 1827 [JLD] &
1898 [HH]; I of Muck 1983 & The
Gobbins 1985 (both *Sublit Surv*).
LONDONDERRY - Moville Buoy 1983
& Portstewart Pt 1983 (both *Sublit Surv*).

Nitophyllum Grev.

Nitophyllum punctatum (Stackh.) Grev.
DA-
Not uncommon, littoral and sublittoral.
Many of the older specimens and records
may be drift. Harvey (*Phycol Brit*)
referred to this species as being
exceedingly abundant and of great size on
the coast of Antrim.
ANTRIM/DOWN - Belfast 1841 [GCH],
pre-1833 (JLD in *Eng Fl*).
DOWN - Lighthouse I drift 1977 [OM];
Donaghadee 1910 (JA 1913); Rolly I drift
1975 [OM]; Doctor's B 1976 (OM);
Ardkeen 1975 [OM det PSD] & 1976
(OM); Granagh B 1976 (OM).
ANTRIM - Giant's Causeway coast pre-
1978 (*Causeway Proj*); Ballycastle pre-
1871 [Miss Hincks]; Murlough B 1891,
1898 [HH] & 1915 [MWR]; Cushendall
pre-1841 (DM in *Harvey's Man*); Red B
1842 [GCH]; near Ballygally Hd 1808
[JT]; nearest coast to The Maiden Rocks

pre-1871 (GD 1871); about 3 miles N of Larne Harbour 1808 [JT], Larne & Larne L 1802, 1803 & 1826 [JLD], dredged 1838 [GCH]; Cairnlough [=Carnlough] B 1836 "I found little of it this season" (JSD 1837).

Phycodrys Kütz.

Phycodrys rubens (L.) Batt.   **DAL**
Common in the lower littoral and sublittoral, also in the drift. Dickie (1871) collected it by dredging at 160m depth off The Maiden Rocks.
DOWN and ANTRIM - Common.
LONDONDERRY - Dredged off Magilligan pre-1948 (HB); Black Sands near Portrush drift 1915 [SW].

Erythroglossum J. Ag.

Erythroglossum laciniatum (Lightf.) Maggs et Hommersand   **DAL**
(*Polyneura laciniata* (Lightf.) Dixon)
(*P. gmelinii* (Lamour.) Kylin)
Very rare in the littoral, however not uncommon in the sublittoral, especially in Co Antrim, to depths of 29m.
ANTRIM/DOWN - Several stations on the NE coast of Ireland pre-1841 (*Harvey's Man*).
DOWN - Near Ballywalter 1982 (*Sublit Surv*); Ballyhenry I 1982 [*Sublit Surv*] & Lee's Wreck 1984 both in Strangford L (*Sublit Surv*); Ardglass 1982 (*Sublit Surv*).
ANTRIM - Several sites on Rathlin I 1983, 1984 & 1985 [*Sublit Surv*] & The Skerries 1982 [*Sublit Surv*], 1984 & 1985 (*Sublit Surv*); Portrush 1984 (*Sublit Surv*) & 1985 [*Sublit Surv*]; Giant's Causeway 1982 [*Sublit Surv*]; Ballintoy 1983 (*Sublit Surv*); Kinbane Hd 1984 (*Sublit Surv*); Fair Hd 1985 (*Sublit Surv*); Murlough B 1898 [HH]; Torr Hd 1983 [*Sublit Surv*]; Red B 1982 [*Sublit Surv*]; nearest coast to The Maiden Rocks pre-1871 (GD 1871); The Maidens 1985 (*Sublit Surv*); Ballygalley Hd 1989 (*NILS*); Larne pre-1830 (JLD in *Alg Brit*); The Gobbins

1985 (*Sublit Surv*).
LONDONDERRY - Portstewart Pt 1985 (*Sublit Surv*).

Polyneura (J. Ag) Kylin.

Polyneura bonnemaisonii (C. Ag.) Maggs et Hommersand   **-A -**
(*P. hilliae* (Grev.) Kylin)
Very rare. Low litttoral rock pools and in the sublittoral.
ANTRIM - Murlough B prob 1898 (HH in *BA* 1902); Larne 1898 [HH].

Radicilingua Papanf.

Radicilingua thysanorhizans (Holm.) Papenf.   **-A-**
Until recently this was considered a 'southern' species only found in SW British Is. It was first recorded in Ireland on the Aran Is, Co Galway in 1981. Between 1982 and 1985 it has been found at several localities in NI at depths between 7 and 29m in Co Antrim.
ANTRIM - Little Skerries 1982 [*Sublit Surv*]; Runkerry Pt 1984 [*Sublit Surv*]; N of Giant's Causeway 1982 [*Sublit Surv*]; Benbane Hd 1984 [*Sublit Surv*]; Rathlin I at: Rue Pt & Bull Pt 1984 (both *Sublit Surv*), White Cliffs 1984 [*Sublit Surv*] & Altachuile B 1985 [*Sublit Surv*]; Fair Hd 1985 (*Sublit Surv*); Murlough B 1985 (*Sublit Surv*); Ringfad 1983 [*Sublit Surv*].

Dasya C. Ag.

Dasya hutchinsiae Harv.   **D—**
Very rare, only one record.
DOWN - Dredged E of Bangor 1838 [JLD det WHH as *Dasya arbuscula* var.].

D. ocellata (Grat.) Harv.   **D—**
Very rare. Only recorded once from NI. A species of S and SW coasts of the British Is (*Sea B I*).
DOWN - Very abundant at Gibbs I 1993 [CAM] F10666.

Heterosiphonia Mont.

Heterosiphonia plumosa (Ellis) Batt.
(Pl.16) **DAL**
Littoral and sublittoral to below 30m.
Epilithic or epiphytic and relatively
common on the more exposed shores of
the British Is.
DOWN - Cultra 1847 [WHH det *"Dasya
coccinea* ß var."]; Barkley Rocks 1984
(*NILS*); Wallaces Rocks 1984 (*NILS*);
Ballyhenry Pt 1980 [CAM]; Kearney
1985 (*NILS*); Granagh B 1985 (*NILS*);
Wreck Port 1985 (*NILS*); Rathmullan Pt
(*NILS*).
ANTRIM - Common.
LONDONDERRY - Magilligan 1838
[GCH]; Downhill Strand 1984 (*NILS*);
Rinagree Pt 1984 (*NILS*); Black Sands
1915 [SW].

Bostrychia Mont.

Bostrychia scorpioides (Huds.) Mont. **D—**
Locally common. Found near high water
in very sheltered localities and probably
overlooked.
DOWN - Probably locally common in
Strangford L - White Rock 1934 (MJL
1937); Ballymorran B c.1934 (MJL
1935b); near Ballywallon c.1935 (MJL
1935a); Ringhaddy Rapids 1987 (*NILS*);
Granagh B 1986 (*NILS*); N Dundrum
Inner B 1985 [OM]; Mill B 1981 [OM];
Greencastle Rocks 1985 (*NILS*).

Brongniartella Bory

Brongniartella byssoides (Good. et
Woodw.) Schmitz **DA-**
Rare, low littoral and sublittoral.
ANTRIM/DOWN - Belfast B undated
[GCH].
DOWN - Cultra 1837 [WT]; Belfast L
1896 [RH]; Portavogie 1915 [MWR];
Newcastle 1835 [WT]; Annalong 1851
[WT].
ANTRIM - Murlough B 1898 [HH];

Larne 1898 [HH].

Chondria C. Ag.

Chondria dasyphylla (Woodw.) C. Ag.
DAL
Rare. Littoral and sublittoral.
ANTRIM/DOWN - Belfast L pre-1830
(JLD in *Alg Brit & Harvey's Man*), 1838
dredged [GCH].
DOWN - Holywood 1837 [WT];
Carlingford L 1984 [*Sublit Surv*].
ANTRIM - Portballintrae 1949 [WRM];
little island S of Carrickfergus 1810 [JT].
LONDONDERRY - Dredged off
Magilligan between 1937-1948 (HB
1951).

Laurencia Lamour.

Laurencia hybrida (DC.) Lenorm. ex
Duby **DAL**
(*L. platycephala* Kütz.)
Not as common as *L. pinnatifida*.
However under-recorded in the literature.
It may be overlooked in mistake for that
species.
DOWN and ANTRIM - Common.
LONDONDERRY - Black Rock 1917
[prob SW], & 1981 [N Allen det OM].

L. obtusa (Huds.) Lamour. **DAL**
Most uncommon. As far as is known
these records are not of epiphytic
specimens although this is how the
species is usually said to be found.
DOWN - Near Holywood drift pre-1825
[JT], others were named by MWR and
confirmed by ADC; about Bangor it was
said to be: "not uncommon, growing in
pools of seawater" c.1836 (JSD 1837).
ANTRIM - Antrim coast N of Belfast L
c.1836 (JSD 1837); Whitehouse shore
1809 [JT].
LONDONDERRY - Portstewart 1836 [?];
Black Rock 1981 [OM].

L. pinnatifida (Huds.) Lamour.   **DAL**
Very common on the shores of Cos Down
and Antrim, however few records from
Co Londonderry. Epilithic from mid and
lower littoral into sublittoral. Often on
steeply sloping rock faces.
DOWN and ANTRIM - Common.
LONDONDERRY - Black Rock 1917
[almost certainly SW]; Rinagree B 1983
(OM) & Rinagree Pt 1984 (*NILS*).

Odonthalia Lyngb.

Odonthalia dentata (L.) Lyngb.   **DAL**
An easily recognised alga of northern
distribution. Common on the shores of
Cos Down and Antrim, however few
records from Co Londonderry. Low
littoral in deep rock pools, to several
metres sublittoral. Usually epilithic and
associated with the more exposed shores.
DOWN and ANTRIM - Common.
LONDONDERRY - Portstewart B 1976
[J Burgoyne].

Polysiphonia Grev.

Polysiphonia atlantica Kapraun et J. Norris
                                DAL
(*Polysiphonia macrocarpa* Harv.)
Common but probably under-recorded
until recently. It is relatively small and
similar to some others. Said to be often
associated with sand.
DOWN and ANTRIM - Common.
LONDONDERRY - Downhill Strand
1984 (*NILS*); Portstewart pre - 1841
(*Harvey's Man*); Rinagree B 1983 [OM]
& Rinagree Pt 1984 (*NILS*).

P. brodiaei (Dillw.) Spreng.   **DAL**
A common species of rock pools of mid
and low littoral on the more exposed
shores. Epilithic and epiphytic.
DOWN - Bangor 1831 (JLD in WT
1836); Wallaces Rocks 1984 (*NILS*);
Mahee I 1985 (*NILS*); Granagh B 1986
(*NILS*); Ardglass 1851 [WT]; Sheepland

Harbour 1985 (*NILS*); Rathmullan Pt
1985 (*NILS*); Dundrum Inner B (N) 1987
(*NILS*); Newcastle 1838 [WT]; S of
Newcastle 1851 [WT]; Wreck Port 1985
(*NILS*).
ANTRIM - Common.
LONDONDERRY - Downhill Strand
1984 (*NILS*); Portstewart 1838 [GCH] &
1981 [OM]; Rinagree Pt 1984 (*NILS*).

P. denudata (Dillw.) Grev. ex Harv.   **D/A-**
Very rare, only one old doubtful record,
neither the collector nor determiner is
known and the location is uncertain.
DOWN or ANTRIM - Mill B (there is a
Mill B in both of these counties), growing
on a buoy rope 27 February 1847 [**BEL**]
F1474.

P. elongata (Huds.) Spreng.   **DAL**
A relatively large species up to 30cm in
length and similar to *P. elongella*. Low
littoral into the sublittoral to a depth of
22m.
DOWN - Holywood poss pre-1867 [poss
GCH det ADC]; Whitechurch 1986
(*NILS*); Rolly I 1975 [OM det YMC];
Mahee I 1985 (*NILS*); Doctor's B 1976
[OM]; Black Neb inlet 1986 (*NILS*);
Robin's Rock 1986 (*NILS*); Portavogie
1988 [OM]; Dunnyneill Is 1982 (*Sublit
Surv*); Ballyhenry B 1986 (*NILS*);
Kearney 1987 [OM]; Quoile Estuary 1982
[*Sublit Surv*]; Granagh B 1976 [OM];
Ardglass B 1985 (*NILS*); Killough 1834
[GCH], 1982 & 1983 [*Sublit Surv*];
Ringboy 1985 (*NILS*); Rathmullan Pt
1985 (*NILS*); Craigalea 1980's (*NILS*);
Dundrum B 1984 [*Sublit Surv*]; Bloody
Br 1851 [WT]; Kilkeel 1983 [*Sublit
Surv*]; Greencastle Rocks 1985 (*NILS*).
ANTRIM - On Rathlin I: Arkhill B 1985,
Cooraghy B 1985 & E coast 1985 (all
Sublit Surv); Portballintrae 1961
[MPHK]; Red B 1982 [*Sublit Surv*];
Carrickfergus pre-1847 [WMcC].
LONDONDERRY - Balls Pt 1987
(*NILS*).

P. elongella Harv.   **DA-**
Very rare. Only a few old records of this
species in NI. Possibly a species of more
exposed conditions.
DOWN - S of Newcastle 1851 [WT].
ANTRIM - Nearest coast to The Maiden
Rocks pre-1871 (GD 1871); Larne pre-
1841 (DM in *Harvey's Man*);
Carrickfergus (*BA* 1874); Belfast L pre-
1841 (JLD & WT in *Harvey's Man*).

P. fibrata (Dillw.) Harv.   **DAL**
Rare. Only recorded from the littoral.
DOWN - Black Neb 1986 (*NILS*);
Portaferry 1837 [WT]; Coney I 1985
(*NILS*); Ringboy 1985 (*NILS*); Dundrum
B 1851 [WT].
ANTRIM - Giant's Causeway 1853 [WS
conf ADC]; Port Vinegar 1985 (*NILS*); I
Magee pre-1841 (DM in *Harvey's Man*).
LONDONDERRY - Black Rock pre-1836
(DM in *Fl Hib*).

P. fibrillosa (Dillw.) Spreng.   **D—**
Extremely rare, only two or three records.
DOWN - Bangor pre-1902 (*Batters' Cat*).
A specimen possibly of this species was
collected at Sandeel B 1972 [OM] F2762;
St John's 1985 (*NILS*).

P. fucoides (Huds.) Grev.   **DAL**
(*P. nigrescens* (Huds.) Grev.)
(*P. violacea* (Roth) Spreng.)
Common growing epilithically in rock
pools of the low littoral and in the
sublittoral to 16m.
DOWN and ANTRIM - Common. First
recorded from these counties at Bangor
as "*Conferva fucoides*" 1809 [JT] F36.
LONDONDERRY - Near McKinney's
Bank & at The Gt Bank in L Foyle 1983
[*Sublit Surv*]; Magilligan 1937-1948 (HB
1951); Downhill Strand 1984 (*NILS*);
Castlerock 1982 [OM].

P. furcellata (C. Ag.) Harv.   **-A-**
Very rare, however possibly under-
recorded.

ANTRIM - All the published records
seem to be based on one single find:
Carrickfergus 1845 [WMcC]. This has
been published several times including:
Nat Prins & Phycol Brit. Possibly the
record of it in Belfast B (*BA* 1874) also
refers to this find.

P. lanosa (L.) Tandy   **DAL**
Very common wherever there is
Ascophyllum nodosum, on which it is
epiphytic, rarely also on *Fucus
vesiculosus*. Littoral on sheltered shores.
DOWN and ANTRIM - Very common
especially in Co Down.
LONDONDERRY - Balls Pt 1987
(*NILS*).

P. nigra (Huds.) Batt.   **DA-**
Apparently rare but probably under-
recorded. Littoral and sublittoral.
DOWN - Bangor 1915 drift [SW];
Ballymacormick Pt 1984 (*NILS*);
Kircubbin Pt 1985 (*NILS*); Yellow Rocks
1986 (*NILS*); Ringhaddy Rapids 1987
(*NILS*); Selk Rock 1986 (*NILS*); Craigalea
1980's (*NILS*); Bloody Br 1985 (*NILS*).
ANTRIM - Portrush 1984 [*Sublit Surv*];
Peggy's Hole 1985 (*NILS*); Port Gorm
1985 (*NILS*); Ballintoy 1983 [*Sublit
Surv*]; The Gobbins 1987 (*NILS*).

P. stricta (Dillw.) Grev.   **DAL**
(*P. urceolata* (Lightf. ex Dillw.) Grev.)
This may be a species complex requiring
further investigation (*See B I*).
A common species of the low-littoral and
sublittoral in Cos Antrim and Down,
however only two records from Co
Londonderry.
DOWN and ANTRIM - Common.
LONDONDERRY - Black Rock pre-1836
(DM in *Fl Hib*); Rinagree Pt 1984 (*NILS*)

P. subulifera (C. Ag.) Harv.   **DA-**
Very rare indeed, no recent records from
NI.
DOWN/ANTRIM - Belfast B pre-1825

(JT in *Phycol Brit*) & 1838 [GCH].
ANTRIM - Carrickfergus pre-1849
[WMcC].

Boergeseniella Kylin

Boergeseniella fruticulosa (Wulf.) Kylin
                                          D-L
(*Polysiphonia fruticulosa* (Wulf.) Spreng.)
   Very rare, however probably under-
   recorded.
   DOWN - "Coast of Down" pre-1841 (WT
   in *Harvey's Man*); Bangor 1835 (WT).
   LONDONDERRY - Black Rock pre-1836
   (DM in *Fl Hib* also *Harvey's Man* &
   others).

B. thuyoides (Harv.) Schmitz   **DA-**
(*Pterosiphonia thuyoides* (Harv.) Batt.)
   Very rare with few recent records from
   NI.
   DOWN - Ardglass 1851 [WT det MWR];
   Greencastle Rocks 1985 (*NILS*).
   ANTRIM - Marie Isla on Rathlin I 1986
   (*NILS*); Portrush B pre-1841 (DM in
   Harvey's Man) & 1985 [WS].

Pterosiphonia Falkenb.

Pterosiphonia complanata (Clem.)
Falkenb.   **DA-**
   Until recently this species had not been
   recorded from NI and was considered a
   southern species reaching only into the
   SW of the British Is (*Prov Atlas*).
   However it has now been recorded from
   NI.
   DOWN - Taggart I 1986 (*NILS*);
   Sheepland Harbour 1985 (*NILS*).
   ANTRIM - Black Hd E 1986 (*NILS*); The
   Burnfoot 1984 (*NILS*).

P. parasitica (Huds.) Falkenb.   **DAL**
   A small species of the low littoral and
   sublittoral. Until recently poorly recorded
   although common especially in the
   sublittoral.
   DOWN and ANTRIM - Common.

LONDONDERRY - Near Rinagree Pt
1983 [OM], NW Rinagree Pt 1984 (*Sublit
Surv*) & two records from NW
Portstewart Pt 1984 (*Sublit Surv*).

Rhodomela C. Ag.

Rhodomela confervoides (Huds.) Silva
                                          DAL
   Common in the low littoral and sublittoral
   to 19m. Epilithic.
   DOWN and ANTRIM - Common, first
   recorded from these counties on the
   Whitehouse shore, Co Antrim pre-1825
   [JT].
   LONDONDERRY - Only one record:
   NW Portstewart Pt 1984 [*Sublit Surv*].

R. lycopodioides (L.) C. Ag.   **DAL**
   Rather rare, however some records noted
   below are worthy of further investigation.
   A northern species of the low littoral and,
   according to the literature, usually
   epiphytic on the stipes of *Laminaria*.
   DOWN - Bangor 1835 (WT 1836), 1836
   [WT] & pre-1841 (WT in *Harvey's Man*)
   & 1853 (*Gifford's Mar Bot*); Ballyhornan
   1982 [OM].
   ANTRIM - Coast of Antrim pre-1830 (Dr
   Scott in *Alg Brit*); Giant's Causeway 1814
   [JT], 1853 & 1856 [WS]; Ballycastle prob
   pre-1871 (prob Miss Hincks in *UCC Cat*);
   Larne pre-1902 (*Batters' Cat & Seaw An
   Co*).
   LONDONDERRY - Portstewart pre-1902
   (*Batters' Cat*); Rinagree Pt 1984 (*NILS*).

Notes

Abbreviations

General

[ ]	Record supported by a voucher specimen
( )	Recorded in the literature or personal observation, not supported by voucher specimens
A	Antrim
agg.	aggregate
anon	anonymous
B	Bay
BEL	Ulster Museum Herbarium
B.N.F.C.	Belfast Naturalists' Field Club
Br	Bridge
c.	*circa*: about or approximately
C	century
cm	centimetre
conf	confirmed by
Co(s)	County, Counties
DoE(NI)	Department of the Environment (Northern Ireland)
D	Down
det	determined by
E	East
F	F followed by a number (e.g. F3650) is an Ulster Museum Herbarium [**BEL**] number referring to a specific specimen
et al.	*et alia*: and others
fl.	flourished
f.	forma
Gd	Grand
Gt	Great
Hd	Head

ie	*id est*: that is
I(s)	Island(s) or Isles
Incl	Including
L	Londonderry
L	Lough
LWM	low water mark
m	metres
mm	millimetres
N	North
NI	Northern Ireland/Northern Irish
Pk	Park
pers comm	personal communication
poss	possibly
pre	before
Prob	probably
Pl.	plate
Pt	Point
p.p.	*pro parte*: in part
qv	*quod vide*: which see
R	River
S	South
sp.	species
ssp.	subspecies
var.	variety
W	West

Collectors, Determiners, Recorders and Authors

All collectors, determiners, recorders and authors cited in this text in an abbreviated form are listed below, followed by those not abbreviated. Published sources are listed separately as references using where necessary the initials below. To each name is added the dates of birth and death where known and in some cases the dates of botanical activity. As the dates of botanical activity may have been deduced they are perforce approximate. The locations of voucher specimens according to the literature and personal knowledge are also noted.

BEL   =Ulster Museum, Belfast.
BM   =Natural History Museum, London.
DBN   =National Botanic Gardens, Glasnevin, Dublin.
K   =Royal Botanic Gardens, Kew.
LINN   =Linnean Society, London
TCD   =Trinity College Dublin, School of Botany.

[ ] = Record supported by a voucher specimen in **BEL** or **BM**. In certain cases of rare or interesting specimens a number beginning with the letter "F" is noted, this is the specific number referring to the specimen stored in the Ulster Museum.

( ) = Records in the literature, or personal observations.

ADC   Cotton, Arthur Disbrowe 1879-1962.
AH   Hazlett, A. fl. 1970's.
BEP   Picton, Bernard E. 1970's onwards. Vouch: **BEL**.
C   Cunningham. Poss R.O. Cunningham 1841-1918.
CAM   Maggs, Christine A. 1970's onwards. Vouch: **BEL**.
CHW   Waddell, Rev. Coslett Herbert 1858-1919. fl. 1890's-1919.
CID   Dickson, C.I.= Meikle, Carola Ivena 1900-1970.
CMH   Howson, C.M. 1970'S onwards. Vouch: **BEL**.
CRH   Hackney, Catherine Ruth née Tyrie. 1974 onwards. Vouch: **BEL**.
CRT   Tyrie, Catherine Ruth see CRH.
CTI   Ingold, C.T. 1920's.
DM   Moore, David 1807-1879. Vouch: **BEL.**
DWC   Connor, D.W. 1980's onwards.
EALB   Batters, Edward Arthur Lionel 1860-1907. Vouch: **BM**.
EG   Getty, Edmund 1799-1857. Vouch: **BEL.**
EMB   Burrows, Elsie M. 1913-1987.
EMH   Holmes, Edward Morrell 1843-1930.
FWK   Knaggs, F.W. 1970's onwards.
G   unidentified.
GCH   Hyndman, George Crawford 1796-1867. Vouch: **BEL**.

GD	Dickie, Prof George 1812-1882. Vouch: **BM, K, LINN**.
GR	Russell. Prof George 1960's onwards.
HB	Blackler, Margaret Constance Helen 1902-1981.
HH	Hanna, Henry fl. 1890-1900's. Vouch: **BEL**.
HMP	Parkes, Hilda M. fl. 1960's-1980's.
IT	Tittley, Ian 1970's onwards.
JA	Adams, John 1872-1950. fl. 1899-1910.
JJR	unidentified. fl. 1915-1916. Vouch: **BEL**.
JLD	Drummond, Dr James Lawson 1783-1853. Vouch: **BEL**.
JMcG	McGurk, J. fl. 1930's.
JMK	Kain, Joanna M. fl 1950's-1970's.
JMW	White, J.M. fl. 1930's.
JP	Price, James H. fl. 1960's-1980's. Vouch: **BEL**.
JPC	Cullinana, John P. fl. 1970's-1978.
JSD	actually JLD q.v.
JT	Templeton, John. 1766-1825. Vouch: **BEL**.
JWW	Wright, J.W. 1930's.
LMI	Irvine, L.M. 1960'S onwards.
FWK	Knaggs, F.W. 1970's onwards.
MC	Clapham, Mabel 1920's-1930's.
MDG	Guiry, Michael D. 1969 onwards.
MHH	Hommersand, Max H.
MPHK	Kertland, Mary Patricia H. 1901-1991 Vouch: **BEL**.
MJL	Lynn, Mary J. 1930's-1960's.
MR	Roberts, Margaret 1960's-1970's.
MWR	Rea, Margaret Williamson. fl. 1915-1916. Vouch: **BEL**.
NFM	McMilligan, Nora F. fl. 1970's-1980's.
OM	Morton, Osborne 1969 onwards. Vouch: **BEL**; **BM**.
PCS	Silva, Paul 1950's onwards.
PH	Hackney, Paul 1969 onwards Vouch: **BEL**.
PSD	Dixon, Peter S. fl. 1950's - 1993. Vouch: **BEL**
RB	Brown, Robert 1773-1858. Vouch: **BM**.
RH	Hensman, R. fl. 1890's. Vouch: **DBN**.
RKB	Brinklow, Richard K. 1970's. Vouch: **BEL**.
RLF	Fletcher, R.L. fl. 1970's onwards.
RLlP	Praeger, Robert Lloyd 1865-1953. fl. 1880-1937. Vouch: **BEL**; **DBN**; **BM**.
RS	Seed, R. fl. 1970's.
RSt	Standen, Richard fl. 1897.
RTW	Wilce, R.T. 1960's onwards
RW	Walker, R. fl. 1980's.
SW	Wear, Sylvanus 1858-1920. fl. 1915-1917. Vouch: **BEL**.
TJ	Johnson, Prof. Thomas 1863-1954. Vouch: **BEL**.
TN	Norton, Trevor A. 1960's onwards.
WHH	Harvey, William Henry. 1811-1866. Vouch: **BEL**; **BM**; **K**; **TCD**.
WH	Hincks, Rev. Prof. William. 1794-1871.
WMcC	McCalla, William 1814-1849. Vouch: **BEL**.
WRM	Megaw, Rev. William Rutledge 1885-1953. Vouch: **BEL**.

WS	Sawers, William 1850's.
WSW	Wright, W.S. 1930's.
WT	Thompson, William 1805-1852. Vouch: **BEL**; **DBN**.
WW	Wade, Walter 1760-1825.
YMC	Chamberlain, Yvonne M. 1960's onwards.

Names not abbreviated:-

N Allen fl. 1981. Vouch: **BEL**.

J Burgoyne fl. 1976. Vouch: **BEL.**

DC Campbell fl. 1880's. Vouch: **BEL**.

Dr D fl 1850's.

Miss Davidson fl. 1830's.

J Doran

M Foslie fl. 1880's-1890's. Vouch: **BEL**.

Mrs Gatty

RN Gregg fl. 1930's.

Hennedy

Miss Hincks fl. pre-1871. Vouch: **BEL**.

AM Irwin c. 1936-1951. Vouch: **BEL**.

CA Johnson

G Johnston

Landsborough poss. Rev. David c. 1826-1912.

Mr Moon fl. 1852. Vouch: **BEL**.

H Murray

Dr Scott fl. pre-1830.

de Valéra, M. 1912-1984.

Topographical Index

All the locations referred to in this flora are listed below with an indication of their county
(**A**, **D** and **L** = Antrim, Down and Londonderry respectively), their grid reference and a note
of their location. A four figure grid reference is given as in a number of cases the location is
of a town or village and the record is obviously from a near-by shore. Some of the sites are
not shown on the 1:50,000 Ordnance Survey maps and are only known locally, while in
other cases there is doubt about their location as there is more than one site with the same, or
similar, name. There is, for example, more than one Black Rock in the north-east of Ireland
not to mention Blackrocks and Black Rocks. As far as possible all the names have been
made consistent in spelling using the Ordnance Survey maps as a basis.

The locations are listed below in alphabetic order and all distances are estimated.

SITE	COUNTY	GRID REF.	LOCATION
		A	
Abbey Rock	D	J5656	Sublittoral in S Strangford Lough.
Altacorry Hd	A	D1552	Headland on N coast of Rathlin Island.
Altachuile B	A	D1352	Bay on the N coast of Rathlin Island.
Annalong	D	J3719	About 5 miles NE of Kilkeel on the S Down coast.
Ardglass	D	J5637	About 10 miles S of Strangford town.
Ardkeen	D	J6056	Over 3 miles N of Portaferry.
Arkill B	A	D1549	Bay on E coast of Rathlin Island.
Arno's Vale	D	J1618	Between Warrenpoint and Rostrevor in Carlingford Lough.
Audley's Castle rocks	D	J5750	In Strangford Lough about 1 mile NW of Strangford town.
Audley's Pt	D	J5751	Rocky point at Audley's Castle.
		B	
Ballintoy	A	D0444	On N Antrim coast about 5 miles NW of Ballycastle.
Balls Pt	L	C6430	On E coast of Lough Foyle near the outflow of the River Roe.
Ballycarry	A	J4494	In Larne Lough about 4 miles NE of Carrickfergus.

9. *Codium* sp. Darragh Island, Strangford Lough, Co. Down, 1987. (Photo. Julia D. Nunn, Ulster Museum.)

10. *Dictyopteris membranacea.* Sublittoral. East of Black Head, Rathlin Island, Co. Antrim, 1989.
(Photo. Bernard E. Picton, Ulster Museum.)

11. Coralline rock pool dominated by *Corallina officinalis* with *Cladophora* sp. on an exposed shore. Portrush, Co. Antrim, 1989. (Photo. G.V. Day.)

12. *Lithothamnion glaciale* from a deep rock pool in the low littoral. Donaghadee, Co. Down, 1984. (Photo. Osborne Morton, Ulster Museum.)

13. *Stenogramme interrupta.* Sublittoral. Lee's Wreck, Strangford Lough, Co. Down, 1984. (Photo. Bernard E. Picton, Ulster Museum.)

14. *Schmitzia hiscockiana*. Loughan Bay near Torr Head, 1983. Collected by Bernard E. Picton, determined by C.M. Howson. (Ulster Museum: F4368.)

15. *Delesseria sanguinea.* Bloody Bridge, Co. Down, 1991. Collected and determined by Osborne Morton. (Ulster Museum: F8949.)

16. *Heterosiphonia plumosa.* Sublittoral. Lee's Wreck, Strangford Lough, Co. Down. (Photo. Bernard E. Picton, Ulster Museum.)

Ballycastle	A	D1241	Town on N Antrim coast.
Ballygalley	A	D3707	About 3 miles NW of Larne.
Ballygalley Hd	A	D3808	Headland about 3 miles NW of Larne.
Ballyhalbert	D	J6463	On E coast of Ards between Ballywalter and Portavogie.
Ballyhenry B	D	J5851	Small bay in Strangford Lough less than 1 mile N of Portaferry.
Ballyhenry I	D	J5752	Small island N of Ballyhenry Bay (q.v.).
Ballyhenry Pt	D	J5751	N point of Ballyhenry Bay (q.v)
Ballyholme	D	J5282	Near Bangor on N coast of Co Down.
Ballyholme B	D	J5182/J5282	Bay E of Bangor on N coast of Co Down.
Ballyhornan	D	J5942 S	Co Down coast 6 miles E of Downpatrick.
Ballymacormick Pt	D	J5284/J5384	E point of Ballyholme Bay near Bangor.
Ballymorran B	D	J5260	Bay near Killinchy on W coast of Strangford Lough.
Ballyquintin Pt	D	J6245	Southern tip of Ards peninsula.
Ballywaddan	D	J5956	About 3 miles N of Portaferry, near Ardkeen.
Ballywallon	D	J5956	Small peninsula near Ardkeen.
Ballywalter	D	J6369	On the E coast of Ards peninsula about 7 miles SE of Donaghadee.
Bann Estuary	L	C——	Between Castlerock and Portstewart.
Bann R	L	C——	River from Lough Neagh flowing through Coleraine to the sea.
Bangor	D	J5-8-	On the N coast of Co Down.
Barkley Rocks	D	J6074	Rocks less than than 2 miles S of Millisle on the Ards peninsula.
Barney's Pt	A	J4598	Small peninsula in Larne Lough.
Bar Hall B	D	J6146	Bay near the tip of the Ards peninsula.
Beddag	A	D1550	On Rathlin Island in Church Bay.
Belfast B or L	D/A	J——	The lough or bay at Belfast.
Benbane Hd	A	C9696	The most northerly part of NI save Rathlin Island.
Black Causeway	D	J5849	Road causeway less than 1 mile SW of Strangford.
Blackcave Tunnel	A	D3905	About 2 miles N of Larne.
Black Hd	A	J4893	S point of Island Magee.
Black Hd on Rathlin I	A	D1050	A headland on the W arm of Rathlin Island near Bull Point.
Black Is	D	J5948	Islands in entrance to Strangford Lough.
Black Neb	D	J5961	Less than 1 mile S of Kircubbin.
Black Neb inlet	D	J5961	Inlet at Black Neb.
Black Rock near Ballywalter			
	D	J6371	Rocks 1.5 miles N of Ballywalter.
Black Rock in Strangford	D	J5455	Small islet 2.5 miles N of Killyleagh.
Black Rock in the Skerries			
	A	C8742	Small island E of Castle Island in the Skerries.

Black Rock	L	C8239	Rocky point between Portstewart and Portrush.
Blackrock Causeway	L	C8239	Causeway at Black Rock in Co Londonderry (q.v.).
Black Rock Sands	L	C8-3-	Probably sands near Black Rock in Co Londonderry (q.v.).
Black Sands	L		Probably refers to Black Rock Sands (see above).
Bloody Br	D	J3826	Bridge over Bloody Bridge River about 3 miles S of Newcastle.
Brown's B	A	D4302	Bay at the N of Island Magee.
Broad Sound	A	C8-4-	The sea between Skerries and Portrush.
Bruce's Cave	A	D1652	NE point of Rathlin Island.
Bull Pt	A	D0850	Western point of Rathlin Island.
Burnfoot, The	A	C9041	Between Portrush and The Giant's Causeway on the N Antrim coast.
Burr Pt	D	J6662	On the Ards peninsula S of Ballyhalbert.
Bushmills	A	C4094	On Bush River in NW Co Antrim.
Bush R outlet	A	C9342	E of Portballintrae in NW Co Antrim.

C

Cairnlough	A		An old spelling for Carnlough (q.v.).
Campbeltown	A	D2818	About half a mile N of Carnlough on the E coast of Co Antrim.
Carlingford L	D	J1-1- & J2-1-	Lough between Cos Down and Louth.
Carnalea	D	J4881	Near Bangor on N coast of Co Down.
Carnlough	A	D2817	E coast Co Antrim 12 miles NE of Larne.
Carnlough B	A	D2816/D2916/D2917	
			Bay at Carnlough on E coast of Co Antrim.
Carnfunnock B	A	D3806	About 3 miles N of Larne.
Carrickarade I	A	D0645	Island about 5 miles NW of Ballycastle on N Antrim Coast. (Carrick-a-Rede).
Carrickfergus	A	J4187	On N shore of Belfast Lough about 10 miles from Belfast.
Carrickmore	A	D1642	Headland near Fair Head on N Antrim coast.
Carricknaford	A	D0345	Headland about 6 miles W of Ballycastle on N Antrim coast.
Carr Rocks, E	A	C8542	Rocks in The Skerries.
Carrstown Pt	D	J6146	Near the S of Ards peninsula.
Castle Chichester B	A	J4792	Bay at Whitehead on N coast of Belfast Lough.
Castle Espie	D	J4967	Near Comber on NW shore of Strangford Lough.
Castle Hd	A	D1452	Near Altacorry on N coast Rathlin Island.

Castle I	D	J5148	Peninsula in S Strangford Lough where Quoile River enters the sea.
Castle I Pt	D	J5359	On causeway between Castle Island (not the Castle I above) & Darragh Island near Whiterock.
Castlerock	L	C7636	Less than 1 mile W of River Bann outflow.
Castleward B	D	J5749/J5849	Bay W of Strangford town.
Causeway coast	A	C9-4-	The Giant's Causeway coast (q.v.).
Cave House & shore	A	D2532	E shore at Cushendun.
Chapel I	D	J5651	Island in Strangford Lough near Greyabbey.
Chapel I causeway	D	J5651	Causeway to Chapel Island.
Chapman's Rock	A	D4601	S of Portmuck on Island Magee.
Church B	A	D1450	Large bay on Rathlin Island.
Clachan Rock	A	D0851	Near Bull Point, W Rathlin Island.
Cloghy Rocks	D	J5947	Rocks at Cloghy in Strangford Lough narrows.
Colin Rock	D	J5556	Sublittoral pladdy in S Strangford Lough.
Colliery B	A	D1441	About 2 miles E of Ballycastle on the N Antrim coast.
Comber R	D	J4-6-	River entering NW Strangford Lough.
Coney I	D	J5536	At Ardglass on S Down coast.
Connswater	D	J3675	River flowing into Victoria Park and thence into Belfast Lough.
Cooraghy B	A	D1050	Bay on S coast of Rathlin Island.
Copeland Is	D	J5-8-	The three islands N of Donaghadee.
Copeland I	D	J5-8-	The largest of the three Copeland Islands.
Cow & Calf	D	J4533	Two islands in Dundrum Bay.
Craigalea	D	J4534	Rocks on N coast of Dundrum Bay.
Craigavad	D	J4281	On N Down coast between Holywood and Bangor.
Craigbain	D	J6371	Rocks on E coast of the Ards peninsula about 2 miles N of Ballywalter.
Cranfield Pt	D	J2710	S point of Co Down at entrance to Carlingford Lough.
Craiglewey	D	J6041	Point about 4 miles S of Strangford.
Crawfordsburn	D	J4682	Point between Helen's Bay and Bangor.
Culkeeragh	L	C4721	E of outflow of River Foyle into Lough Foyle.
Culloden Hotel	D	J4181	On S coast of Belfast Lough between Belfast and Bangor.
Culmore	L	C4622	W of River Foyle outflow about 4 miles N of Londonderry City.
Cultra	D	J4080	On S coast of Belfast Lough between Belfast and Bangor.
Curran Pt	A	D4101	Point within Larne Lough at Larne.

Cushendall	A	D2427	E coast of Co Antrim between Cushendun and Garron Point.
Cushendun	A	D2432	On NE coast of Co Antrim between Torr Head and Cushendall.
Cushendun B	A	D2533/D2532	Bay at Cushendun (above).

D

Dalaradia & Pt	A	J4399	Point on W coast of Larne Lough.
Derginan Pt & B	A	D0952	Point and Bay on W arm of Rathlin Island.
Doctor's B	D	J5962	Bay S of Kircubbin on E coast of Strangford Lough.
Donaghadee	D	J5980	NE coast of Co Down 5 miles E of Bangor.
Donaghadee Sound	D	J5-8-	Sea between Copeland Islands and Donaghadee.
Doon B & Pt	A	D1548	E coast Rathlin Island.
Dorn, The	D	J5956	Narrow inlet near Ardkeen in Strangford Lough.
Downhill	L	C7536	Near Castlerock on N coast of Co Londonderry.
Downhill Strand	L	C——	Shore at Downhill.
Down Rock	D	J6347	Near S point of the Ards Peninsula.
Drummond I	D	J5560	One of the islands of Strangford Lough.
Drumnagreagh Port	A	D3413	Between Glenarm and Larne on E coast of Co Antrim.
Dundrum	D	J4036	In Dundrum Inner Bay, S Co Down.
Dundrum B	D	J4-3-	Large bay in S Co Down.
Dundrum Inner B	D	J4-3-	Smaller sheltered bay off Dundrum Bay.
Dunimeny Castle	A	D1141	W point of Ballycastle Bay.
Dunluce	A	C9041	About 3 miles E of Portrush on N Antrim coast.
Dunluce Castle	A	C9041	At Dunluce (q.v.).
Dunnyneill Is	D	J5453	Two islands in S Strangford Lough about 3 miles NW of Portaferry.
Dutchman, The	A	D1451	In Church Bay on Rathlin Island.

E

E coast	A	D——	The E coast of Rathlin Island.
Eglinton	L	C5220	6 miles NE of Londonderry on the Muff River.

F

Fair Hd	A	D1843	Also known as Benmore. Headland 4 miles NE of Ballycastle.
Farganlack Pt	A	D1052	Northern point on Rathlin Island.

Foreland Pt	D	J5881	Rocky point between Groomsport and Donaghadee.
Foyle, L	A		The lough between Cos Donegal and Londonderry.

G

Ganaway Burn	D	J5262	River entering Strangford Lough N of Whiterock.
Garron Pt	A	D3024	E coast of Co Antrim between Cushendall and Carnlough.
Garron Harbour	A	D3023	S of Garron Point (q.v.).
Giant's Causeway	A	C9-4-	The well known coast between Portballintrae and Benbane Head.
Gibbs I	D	J5149	Small island about 2 miles SW of Killyleagh.
Gid Pt	A	D0044	W point of White Park Bay on the N Antrim coast.
Glenariff R	A	D2425	River entering sea near Cushendall.
Glasdrumman Port	D	J3822	Between Newcastle and Annalong.
Glenarm	A	D3115	Between Carnlough and Ballygalley.
Glenarm B	A	D3015/D3115	
			Bay at Glenarm (q.v.).
Gobbins, The	A	J4-9-	Cliffs on the E coast of Island Magee.
Granagh B	D	J6048	Bay on the Ards peninsula at the Strangford L narrows.
Gd Causeway	A	C9444	Headland in Giant's Causeway.
Gt Bank, The	L	C5430	Sublittoral bank in S Lough Foyle.
Gt Stookan	A	C9444	Headland in Giant's Causeway.
Greenisland	A	J3884	N coast of Belfast Lough, 7 miles N of Belfast.
Green I	D	J5351	Small Island in SW Strangford Lough.
Green I in Carlingford L			
	D	J2411	Small island at entrance to Carlingford Lough.
Greencastle	D	J2411	At entrance to Carlingford Lough.
Greencastle Rocks	D	J2311	Rocky point at entrance to Carlingford Lough.
Greenore Pt	D	J2211	The point itself is at the entrance to Carlingford Lough in Co Louth. However as the species found were recorded N of Greenore Point it is included here.
Groomsport	D	J5383	2 miles NE of Bangor.

H

Halfway House	A	D3608	Between Glenarm and Ballygalley on E coast of Co Antrim.

Helen's B	D	J4682	Between Belfast and Bangor in S Belfast Lough.
Herring B	D	J5865	Near Greyabbey on the NE coast of Strangford Lough.
Highlandman's Bonnet, The	A	C4444	Bay in Giant's Causeway.
Highlandman Rock	A	D4514	A northern rock of The Maidens.
Holm B	D	J5353	At Killyleagh on the E shore of Strangford Lough.
Holywood	D	J3979	S shore of Belfast Lough.
Hood's Ferry	A		On Island Magee opposite Larne.
Horse I	D	J5960	Small island near Kircubbin in N Strangford Lough.

I

Isle of Magee	A		Island Magee. Large peninsula in Co Antrim enclosing Larne Lough.
Isle O'Valla	D	J5948	Small island in Strangford Lough narrows.

K

Kearney	D	J6551	On the SE shore of the Ards Peninsula.
Kearney Pt	D	J6451	Near Kearney on the E shore of Ards Peninsula.
Kilclief	D	J5945	Near entrance to Strangford Lough 2.5 miles S of Strangford town.
Kilclief Pt	D	J5946	Rocky point near Kilclief.
Kilkeel	D	J3113	Near the most southerly point of Co Down.
Killard Pt	D	J6143	Point 4 miles S of Strangford town.
Killough	D	J5336	About 6 miles SE of Downpatrick.
Killough B	D	J5-3-	Bay at Killough (q.v.).
Killough Harbour	D	J5-3-	Harbour at Killough (q.v.).
Killowen	D	J1815	N shore of Carlingford Lough.
Killowen Bank	D	J1915	Sand bank near Killowen.
Killeany	A	D1251	About halfway along W arm of Rathlin Island.
Killinchy	D	J4960	Between Comber and Killyleagh.
Killyleagh	D	J5252	On W shore in S Strangford Lough.
Kinbane Hd	A	D0843	N coast of Co Antrim about 3 miles W of Ballycastle. Also known as White Head.
Kinnegar & **Kinnegar Rocks**	D	J5977	Rocky point on Ards peninsula between Donaghadee and Millisle.
Kinrea	A	D1153	On N shore of Rathlin Island.
Kircubbin	D	J5962	NE shore of Strangford Lough.

Kircubbin B	D	J5963	Bay at Kircubbin (q.v.).
Kircubbin Pt	D	J5962	Point at Kircubbin (q.v.).

L

Ladle, The	A	C9343	Part of Giant's Causeway.
Larne	A	D4003	At mouth of Larne Lough.
Larne harbour	A	D4103	Harbour at Larne.
Larne L	A		Lough S of Larne.
Larry Bane B	A	D0544	Bay near Ballintoy on N Antrim coast between Benbane Head and Kinbane.
Larry Bane Hd	A	D0445	Headland at Ballintoy on N Antrim coast.
Layd Church	A	D2429	On E Antrim coast between Cushendun and Cushendall.
Lee's Wreck	D	J5752	S Strangford Lough near Portaferry.
Lagan	D/A	J——	River entering Belfast Lough at Belfast.
Legnaboe	D	J5840	Between Ballyhornan and Ardglass on S Down coast.
Lighthouse I	D	J5985	The NW island of the three Copeland Islands about 4 miles N of Donaghadee.
Lighthouse (W)	A	D0951	W Rathlin Island.
Lighthouse platform (E)	A	D1652	NE point of Rathlin Island.
Limestone Rock	D	D5555	In S Strangford Lough about 2.5 miles NE of Killyleagh.
Little Skerries	A	C8542/C8642	
			Islands N of Portrush.
Lower Doaghs	L	D6637	Near Magilligan Point in N Lough Foyle.
Long Sheelah	D	J5658	Pladdy in S Strangford Lough, 4 miles NE of Killyleagh.
Loughan B	A	D2437/D2438	
			Bay on E Antrim coast between Cushendun and Torr Head.
Loughshore Pk	A	J3683	Between Belfast and Carrickfergus on the N shore of Belfast Lough.
Luke's Pt	D	J5182	Between Bangor Bay and Ballyholme Bay on N Co Down coast.

M

McIlroy's Port	A	J4502	Near Portmuck, N Island Magee.
McKinney's Bank	L	C6337	Sublittoral bank in N Lough Foyle.
McKinney's Buoy	L	C6337	N Lough Foyle.
Macedon Pt	A	J3581	Small point on coast of N Belfast Lough.
Magheramorne	A	J4398	W coast of Larne Lough about 3 miles SE of Larne.
Magee I	A		Large peninsula in S Co Antrim enclosing Larne Lough.

Magilligan & Magilligan shore

	L	C——	Strand and point on N Londonderry coast.
Mahee I	D	J5-6-	One of the larger islands in N Strangford Lough.

Maiden Rocks, The or The Maidens

	A	D4-1-	Group of small islands in N Channel, 7 miles NE of Larne.
Marchburn Port	A	J4894	SE of Island Magee less than 1 mile N of Black Head.
Marie Isla	A	D1352	About halfway along N coast of Rathlin Island.
Marlfield B	D	J5753	Small bay about 2 miles N of Portaferry.
Marlfield Rocks	D	J5753	Rocks at Marlfield Bay (above).
Mew I	D	J6086	The NE island of the three Copeland Islands (q.v.).
Middle Bank in Larne L	A	D4200	Sublittoral bank in N Larne Lough.
Middle Bank	L	C5627	Sublittoral bank in S Lough Foyle.
Mid I B	D	J5959	About 2 miles S of Kircubbin in Strangford Lough.
Mill B	D	J2-1-	Large bay on the N shore of Carlingford Lough.
Mill Quarter B	D	J6043	S Co Down coast about 3 miles S of Strangford.
Mountstewart	D	J5570	NE shore of Strangford Lough.
Moville Buoy	L	C6137	N Lough Foyle.
Muck, I of	A	D4602	Small island on E coast of Island Magee.
Mullartown Pt	D	J4021	Small point 1 mile N of Annalong.
Murlough/Murlough B	A	D1941/D2041	
			About halfway between Fair Head and Torr Head on N Co Antrim coast.
Murphy's Pt	D	J3618	About 1 mile SW of Annalong.

N

New Br	L	C4519	S Lough Foyle.
Newcastle	D	J3731	Town in S Co Down.
Newmill	A	J4695	In S Larne Lough.
North Rocks	D	J6756	Rocks about 2 miles offshore E of Cloghy.
North B Portrush	A	J8-4-	Bay at Portrush.
Nicholson's Pt	D	J2911	Small point S of Kilkeel.

O

Oldmill B	A	J4596	Small bay in S Larne Lough.
Old Pier	A	D2524	S of Red Bay.
Orlock Pt	D	J5683	N Co Down coast between Groomsport and Donaghadee.
Oweydoo	A	D1448	On S arm of Rathlin Island.

P

Pawle I	D	J5457	Island in Strangford Lough 3 miles N of Killyleagh.
Peggy's Hole	A	C9142	Less than 1 mile W of Portballintrae.
Portaferry	D	J5950	In Strangford Lough narrows.
Portandoon	A	D0951	Near W point of Rathlin Island.
Portavogie	D	J6659	About 7 miles NE of Portaferry on E coast of Ards Peninsula.
Portawillin	A	D1651	E coast of Rathlin Island.
Portbradden	A	J0044	In White Park Bay on N Co Antrim coast.
Portballintrae	A	C9242	On N Co Antrim coast between Portrush and Giant's Causeway.
Portdoo	A	D1942 N	Co Antrim coast SE of Benmore Head.
Port Gorm	A	C9142	Near Portballintrae on N Antrim coast.
Port Moon & B	A	C9745	Near Benbane Head on N Antrim coast.
Portmore	A	D2537	Between Torr Head and Cushendun on E Co Antrim coast.
Port-na-Tober & Hd	A	C9645	Near Benbane Head on N Co Antrim coast.
Portmuck	A	D4502	On NE coast of Island Magee.
Portrush & B	A	C8540	N Antrim coast between Portstewart and Portballintrae.
Portstewart, B & Pt	L	C8138	Between Castlerock and Portrush.
Port Vinegar	A	D2428	Near Cushendall on the E Antrim coast.

Q

Quarterland B	D	J5258/J5358	Bay in Strangford Lough about 3.5 miles N of Killyleagh.
Quintin B	D	J6350	Bay on E coast of Ards Peninsula about 2 miles from Portaferry.
Quoile Estuary	D	J5-5-	Estuary to Quoile River which enters Strangford Lough S of Killeagh.

R

Ramore Hd	A	C8541	Headland at Portrush.
Rathlin I	A	C——	Large island N of Ballycastle.
Rathmullan Pt	D	J4835	N coast of Dundrum Bay.
Red Arch	A	D2426	Near Cushendall on E Antrim coast.
Red B	A	D2-2-	Near Cushendall on E Antrim coast.
Riding Stone	A	D4602	NE coast of Island Magee.
Rinagree, B & Pt	L	C8339	Coast between Portstewart and Portrush.
Ringbane	D	J5349	S Strangford Lough about 2 miles S of Killyleagh.
Ringboy	D	J4935	In Dundrum Bay about 2.5 miles W of Killough.

Rinburr Pt	D	J5755	In Strangford Lough on W coast of Ards Peninsula.
Ringfad	D	J5535	Rockypoint 1 mile S of Ardglass.
Ringfad	A	D2920	Headland about 2.5 miles S of Garron Point.
Ringhaddy Rapids	D	J5358	In Strangford Lough between Killinchy and Killyleagh.
Ringhaddy Sound	D	J5459	Deep water in Strangford Lough W of Ringhaddy.
Ringneill Br	D	J5-6-	W shore of Strangford Lough about 3 miles SE of Comber.
Robin's Rock	D	J6661	Near Portavogie on E coast of Ards Peninsula.
Rockport	D	J4382	Between Holywood and Bangor on S shore of Belfast Lough.
Rock Ryan	A	C?	Near Portrush. (Not shown on OS maps, location not determined.)
Roe Br	L	C6429	Railway bridge over River Roe on E coast of Lough Foyle.
Roe I	D	J5460/J5461	Island in W Strangford Lough about 2 miles from Whiterock.
Rolly I	D	J5-6-	Island in Strangford Lough about 4.5 miles SE of Comber.
Rostrevor	D	J1718	N shore of Carlingford Lough.
Rough I	D	J4968	N Strangford Lough about 2 miles E of Comber.
Ruebane Pt	A	D2041	NE Co Antrim between Fair Head and Torr Head.
Runkerry Pt	A	J9343	Headland between Bushmills and Giant's Causeway on N Co Antrim coast.
Rue Pt	A	D1547	S headland of Rathlin Island.
Rue Pt	D	J5949	In Strangford Lough narrows.
Runkerry Pt	A	J9343	Between Portballintrae and Giant's Causeway on N Antrim coast.

S

St John's Pt	D	J5233	Headland forming E point of Dundrum Bay.
Saltwater Br	D	J6059	Bridge over Blackstaff River where it enters Strangford Lough.
Sandeel B	D	J5583	Small bay on N Down coast between Bangor and Donaghadee.
Seacourt	A	D4004	Rocks less than 1 mile N of Larne
Selk Rock	D	J5752	Island on SE shore of Strangford Lough.
Sheepland Harbour	D	J5838	About 2 miles N of Ardglass in S Down
Skerries, The	A	C8-4-	Group of small islands about 1 mile N of Portrush.

Sketrick I	D	J5262	Island on E coast of Strangford Lough.
Slidderycove Pt	A	C8840	N Antrim coast 2 miles E of Portrush.
South I	D	J5666	Near Greyabbey in N Strangford Lough.
Springvale	D	J6-6-?	Believed to be near Ballywalter.
Stackamore	A	D1552	At Altacorry on N coast of Rathlin Island.
Stalka Rock	D	J2-1-	In Carlingford Lough.
Storks, The	A	C8942	Island E of The Skerries.
Straidkilly Pt	A	D3016	Headland between Carnlough and Glenarm on E Antrim coast.
Strangford Harbour	D	J5849	Harbour at Strangford town.
Strangford L	D	J——	Large Lough in Co Down.
Strangford L narrows	D	J——	Entrance to Strangford Lough where there is a fast race of water at ebb and flow of tide.
Swilly L	**Co Donegal**		A lough not in NI but referred to in this work.
Swineley B & Pt	D	J4782	A bay and point less than 2 miles W of Bangor.

T

Taggart I	D	J5354/J5355	Large island in Strangford Lough, N of Killyleagh.
Tara Pt	D	J6-4-	Near S of Ards peninsula.
Thompson's Bank	A/D	J——	N coast of Belfast Lough (location uncertain).
Torr Hd	A	D2340	NE headland of Co Antrim.
Turnly's Port	A	D2921	About halfway between Garron Point and Carnlough on the Co Antrim coast.

U

Ushet Pt	A	D1547	Near the S headland of Rathlin Island.

V

Vidal Rock	D?	J2509	At entrance to Carlingford Lough possibly on the Co Louth side.

W

Wallaces Rocks	D	J6467	Rocky outcrop on E coast of Ards Peninsula near Ballywalter.
Warrenpoint	D	J1418	Town in NW Carlingford Lough.
Wash Tub	A	C8540/C8541	At or near Portrush.
Watson Rocks	-	J2110	In Carlingford Lough, probably on the Co Louth side but noted here.

Wee Pill	D	J5636	Just S of Ardglass.
Whitebay Pt	A	D3215	E coast of Co Antrim near Glenarm.
Whitechurch	D	J6370	E coast of Ards peninsula about 1 mile N of Ballywalter.
White Cliffs	A	D0950	S facing cliffs near Bull Point on Rathlin Island.
White Hd	A	J4791	Headland S of Island Magee about 13 miles NE of Belfast. (Kinbane Head, near Ballycastle, is also known as White Head. All the White Head records noted here are from the White Head NE of Belfast).
Whitehouse	A	J3480	Northern outskirts of Belfast.
Whitehouse Pt	A	J3480	Presumed to be the point at Whitehouse.
White Lady	A	D2924	On E Antrim coast near Garron Point.
White Pk B	A	D0144/D0244	
			Bay on N Co Antrim coast a few miles E of the Giant's Causeway.
White Rock	D	J5261	On W coast of Strangford Lough near Killinchy.
White Rocks	A	C8840	N Antrim coast about 2 miles E of Portrush.
Wilson's Pt	D	J4982	W Point of Bangor Bay.
Wreck Port	D	J3618	About 1 mile S of Annalong on S Down coast.

Y

Yellow Rocks	D	J5957	E side of Strangford Lough about 4 miles N of Portaferry.

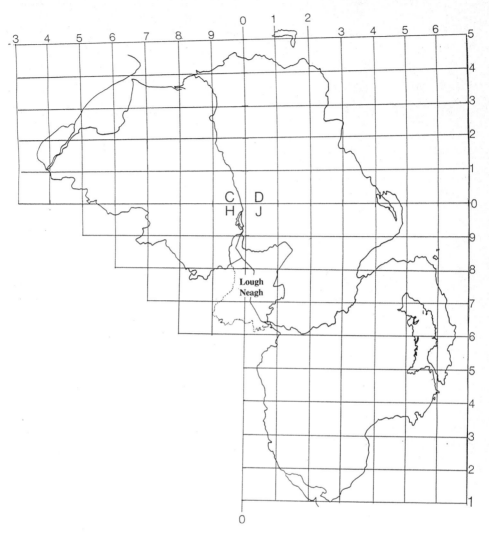

Fig. 2. The 10km x 10km squares of the Irish Grid in the three Northern Irish counties with a coastline. (Hackney, 1991).

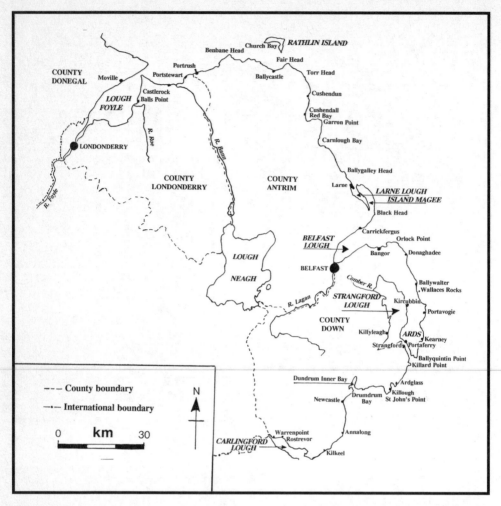

Fig. 3. Location of some places listed in the Topographical Index (after Hackney, 1991).

References
I. Abbreviated

A & A   Adey, W. H. and Adey, P. J. 1973. Studies on the
                        biosystematics and ecology of the epilithic crustose Corallinaceae
                        of the British Isles. *Br. phycol. J.* **8***:* 343 - 407.
Alg Brit   Greville, R. K. 1830. *Algae Britannicae,...* Edinburgh.
Batters' Cat   Batters, E. A. L. 1902. A catalogue of the British marine algae.
                        J. Bot., Lond. **40** (Suppl.): 1-107.
BA 1874   Anon. 1874. *Guide to Belfast and the Adjacent Counties. By*
                        Members of the Belfast Naturalists' Field Club. Belfast.
BA 1902   H.(anna), H., 1902. Algae - marine algae. pp. 144 - 146. *In:*
                        Hanna, H., Praeger, R. Ll., and Waddell, C. H., Botany. *In:*
                        Bigger, F. J., Praeger, R. Ll. and Vinycomb, J. *A Guide to*
                        Belfast and the Counties of Down and Antrim. Prepared for the
                        Meeting of the British Association by the Belfast Naturalists'
                        Field Club. Belfast.
Br Check-list   Parke, M. and Dixon, P. S. 1976. Check-list of British marine
                        algae - third revision. *J. mar. biol. Ass. U.K* **56***:* 527 - 594.
Brit Seaweeds   Landsborough, D. 1857. A *Popular History of British Sea*
                        weeds, ... Third edition. London.
Causeway Proj   Anon. 1978. *Giant's Causeway Environmental Studies Project.*
                        1975. Revised and enlarged 1978. The Teachers Centre, New
                        University of Ulster.
Clare I S   Cotton, A. D. 1912. Marine Algae. *In:* Praeger, R. Ll. A
                        biological survey of Clare Island in the County of Mayo,
                        Ireland and of the adjoining district. *Proc. R Ir. Acad.* **31** sect. I
                        (15): 1-178.
Eng Bot   Smith, J. E. and Sowerby, J. 1790 - 1814. *English Botany; or*
                        Coloured Figures of British Plants their Essential Characters,...
                        1 - 36. London.
Eng Fl   Hooker, W. J. 1833. *The English Flora of Sir James Edward*
                        Smith. Class xxiv Cryptogamia. London. **5**: (1).
Fl Hib   Harvey, W. H. and Mackay, J. T. 1836. Flora Hibernica.
                        part Third. Algae. pp.157 - 254. *In:* Mackay, J. T., *Flora*
                        Hibernica... **2***:* pp.279 +[1]. Dublin.
Gifford's Mar Bot   Gifford, I. 1853. *The Marine Botanist;* Brighton and London.
Harvey's Man   Harvey, W. H. 1841. A *Manual of the British Algae:* John
                        Van Voorst, London.
NA Checklist   South, G.R. and Tittley, I. 1986. *A Checklist and Distributional*
                        Index of the Benthic Marine Algae of the North Atlantic Ocean.
                        Huntsman Marine Laboratory and British Museum (Natural
                        History). St Andrews and London.
Nat Prins   Johnstone, W. G. and Croall, A. 1859. *The Nature - printed*

British Sea-weeds... **1** and **2**. London.

Johnstone, W. G. and Croall, A. 1860. *The Nature - printed British Sea-weeds ...* **3** and **4**. London.

NILS   *Northern Ireland Littoral Survey.* 1988. (Institute of Offshore Engineering, Heriot-Watt University). Un-published report for DoE(NI). Data in Ulster Museum.

[Unless indicated to the contrary all *NILS* records are not supported by voucher specimens.]

Ord S L   (Moore, D.), 1837 (?). Botany, pp 6 -13. *In:* Anon., Ordnance Survey. Notices. pp. 1-16. *In:* Larcom, T.A. *Ordnance Survey of the County of Londonderry.* **1***:* Dublin.

Phycol Brit   Harvey, W. H. 1846-51. *Phycologia Britannica:* **1**- **3**. London.

Prov Atlas   Norton, T. A. (Ed.) 1985. *Provisional Atlas of the Marine Algae of Britain and Ireland.* NERC. Huntingdon.

Railway Guide   Praeger, R. Ll. 1898. *Belfast and County Down Railway Company. Official Guide to County Down and the Mourne Mountains with Seventy Photographs of Scenery by R. Welch, Belfast, Maps and Other Illustrations.* Belfast .

Rev Clado   van den Hoek, C. 1963. *Revision of the European Species of Cladophora.* Leiden.

Sea B I   Dixon, P. S. and Irvine, L. M. 1977. *Seaweeds of the British Isles.* **1**. *Rhodophyta, Part 1. Introduction, Nemaliales, Gigartinales.* British Museum (Natural History), London.

Irvine, L. M. 1983. *Seaweeds of the British Isles.* **1.** *Rhodophyta, Part 2A. Cryptonemiales (sensu stricto), Palmariales, Rhodymeniales.* British Museum (Natural History), London.

Fletcher, R. L. 1987. *Seaweeds of the British Isles.* **3**. *Fucophyceae (Phaeophyceae), Part 1.* British Museum (Natural History, London.

Burrows, E. M. 1991. *Seaweeds of the British Isles.* **2**. *Chlorophyta.* Natural History Museum, London.

Maggs, C.A. and Hommersand, M.A. 1993. *Seaweeds of the British Isles.* **1**. *Rhodophyta, Part 3A Ceramiales.* HMSO.

Seaw An Co   Adams, J. 1907. The seaweeds of the Antrim coast. *Scient. Pap. Ulster Fish. Biol. Ass.,* 1(1): Paper iv.

Sublit Surv   Erwin, D. G., Picton, B. E., Connor, D. W., Howson, C. M., Gilleece, P. and Bogues, M. J. 1986. *The Northern Ireland Sublittoral Survey.* Ulster Museum. Un-published report for DoE(NI). Data in Ulster Museum.

Syn Seaw   Anon and Harvey, W. H. 1857. *Synopsis of British Seaweeds Compiled from Professor Harvey's Phycologia Britannica.* London.

UCC Cat   Parkes, H. M. 1953. *Catalogue of Marine Algae housed in The Botany Department, University College, Cork.* Unpublished.

II. Other References

ADAMS, J. 1904a. *Chantransia alariae* Jónss. in the British Isles. pp. 351-352. *In,* ANON, Short Notes. *J. Bot. Lond.* **42**: 351-354.

ADAMS, J. 1904b. Note on some seaweeds occurring on the Antrim coast. *Ir. Nat.* **13**: 138.

ADAMS, J. 1908. A synopsis of Irish algae, freshwater and marine. *Proc. R. Ir. Acad.,* **27**B: 11- 60.

ADAMS, J. 1913. Some new localities for marine algae. *Ir. Nat.* **22**: 12-13.

ANON 1899. Dublin Microscopical Club. pp. 105-106. *In* ANON. Proceedings of the Irish Societies. *Ir. Nat.* **8**: 105-108.

BATTERS, E.A.L. 1897. New or critical British marine algae. *J. Bot. Lond.* **35**: 433-440.

BLACKLER, H. 1937. The alga *Colpomenia sinuosa* Derb. et Sol. in Ireland. *Ir. Nat. J.* **6**: 196-197.

BLACKLER, H. 1951. An algal survey of Lough Foyle, North Ireland. *Proc. R. Ir. Acad.* **54**B: 97-139.

BLACKLER, H. & McMILLAN, N. F. 1953. Some interesting marine mollusca and algae from Port Ballintrae, in North Ireland. *Ir. Nat. J.* **11**: 76.

BLIDING, C. 1963. A critical survey of European taxa in Ulvales. Part 1 *Capsosiphon, Percursaria, Blidingia, Enteromorpha. Op. Bot. Soc. Bot. Lund* **8**: 1-160.

CHAMBERLAIN, Y. M. 1983. Studies in the Corallinaceae with special reference to *Fosliella* and *Pneophyllum* in the British Isles. *Bull. Br. Mus. nat. Hist.* (Bot) **11**: 291-463.

CHAMBERLAIN, Y. M. 1990. The genus *Leptophytum* (Rhodophyta, Corallinaceae) in the British Isles with descriptions of *Leptophytum bornetii, L. elatum* sp. nov. and *L. laeve. Br. phycol. J.* **25**: 179-192.

CHAMBERLAIN, Y. M. 1991. Historical and taxonomic studies in the genus *Titanoderma* (Rhodophyta, Corallinales) in the British Isles. *Bull. Br. Mus. nat. Hist.* (Bot.) **21** (1): 1-80.

CHAMBERLAIN, Y. M., IRVINE, L. M. & WALKER, R. 1988. A redescription of *Lithophyllum crouanii* (Rhodophyta, Corallinales) in the British Isles with an assessment of its relationship to *L. orbiculatum. Br. phycol. J.* **23**: 177-192.

CHAMBERLAIN, Y. M, IRVINE, L. M. & WALKER, R. 1991. A redescription of *Lithophyllum orbiculatum* (Rhodophyta, Corallinales) in the British Isles and a reassessment of generic delimitation in the Lithophylloideae. *Br. phycol. J.* **26**: 149-167.

CLAPHAM, M. 1926. The ecology of rock pools. *Ir. Nat. J.* **1**: 48-51.

CLAPHAM, M. 1930. Botanical Society of Northern Ireland. Coastal survey no.11. Ballyholme Bay to Carnalea. *Ir. Nat. J.* **3**: 56-64.

CULLINANE, J. P. 1978. A preliminary account of the distribution of *Cordylecladia erecta* (Grev.) J. G. Ag. (Rhodophyta: Rhodymeniales) in Ireland and the British Isles. *Scient. Proc. R. Dubl. Soc.* **6** (Series A), 5: 49-58.

DE VALÉRA, M. 1962. Some aspects of the problem of the distribution of *Bifurcaria bifurcata* (Valley) Ross on the shores of Ireland, north of the Shannon Estuary. *Proc. R. Ir. Acad.* **62**B: 77-100.

DICKIE, G. 1871. Notes on the distribution of algae. *J. Bot. Lond.* **9**: 70-72.

DIXON, P. S. 1960. Studies on marine algae of the British Isles: the genus *Ceramium*. *J. mar. biol. Ass. U.K.* **39**: 331-374.

DIXON, P. S. & PRICE, J. H. 1981. The genus *Callithamnion* (Rhodophyta: Ceramiaceae) in the British Isles. *Bull. Br. Mus. nat. Hist.(Bot)* **9**: 99-141.

DRUMMOND, J. S. (Actually J. L. Drummond) 1837. Directions for the preservation of sea plants, with miscellaneous remarks on a number of species collected at Cairnlough Bay, on the coast of Antrim, in the months of May and June 1836. *Mag. Zool. Bot.* **2**: 144-158.

EDWARDS, P. 1975. Evidence for a relationship between the genera *Rosenvingiella* and *Prasiola* (Chlorophyta). *Br. phycol. J.* **10**: 291-297.

FOSLIE, M. 1898. Some new or critical Lithothamnia. *K. norske Vidensk. Selsk. Skr.* **1898** No.6: 1-19.

FOSLIE, M.1905. Remarks on northern Lithothamnia. *K. norske Vidensk. Selsk. Skr.* **1905** No.3: 1-138.

GUIRY, M.D. 1978. A concensus and bibliography of Irish seaweeds. *Bibliotheca Phycologia* **44**: 1-287.

GUIRY, M. D., IRVINE, L. M. & MORTON, O. 1981. Notes on Irish marine algae - 4. *Gymnogongrus devoniensis* (Greville) Schotter (Rhodophyta). *Ir. Nat. J.* **20**: 288-292.

HANNA, H. 1899. Some algae from the Antrim coast. *Ir. Nat.* **8**: 155-156.

HARVEY, W. H. 1857. Short descriptions of some new British algae with two plates. *Nat. Hist. Rev.* **4**: 201-204.

HAZLETT, A. & SEED, R., 1976. A study of *Fucus spiralis* and its associated fauna in Strangford Lough, Co. Down. *Proc. R. Ir. Acad.,* **76**B: 607-618.

INGOLD, C. T. 1929. Botanical Society of Northern Ireland - Coastal Survey: 1. Grasswrack Community in Ballyholme Bay. *Ir. Nat. J.* **2**: 165-166.

JOHNSON, T. & HENSMAN, R. 1896. Algae from the north side of Belfast Lough. (Dredged by the B.N.F.C. Expedition, 4th July, 1896.) *Ir. Nat. J.* **15**: 252-253.

KERTLAND, M.P.H. 1966. Bi-centenary of the birth of John Templeton. A.L.S. 1766 - 1825 (with a note on the discovery of part of his herbarium). *Ir. Nat. J.* **15**: 229-231.

LARSEN, J. 1981. Crossing experiments with *Enteromorpha intestinalis* and *E. compressa* from different European localities. *Nord. J. Bot.* **1**: 128-136.

LYNN, M. J. 1935a. Rare algae from Strangford Lough. - Part I. *Ir. Nat. J.* **5**: 201-208.

LYNN, M. J. 1935b. Rare algae from Strangford Lough. - Part II. *Ir. Nat. J.* **5**: 275-283.

LYNN, M. J. 1937. Notes on the algae of the district of Whiterock, Strangford Lough. *Ir. Nat. J.* **6**: 192-195.

LYNN, M. J. 1949. A rare alga from Larne Lough. *Ir. Nat. J.* **9**: 301-304.

LYNN, M. J. 1952. Algae. pp.58-59. *In*: SMALL, J., Botany. *In*: JONES, E., *Belfast in its Regional Setting. A Scientific Survey Prepared for the Meeting held in Belfast 3rd to 10th September, 1952*.pp. 211. Belfast: British Association.

LYNN, M. J. 1960a. Coastal Survey X (New Series) Southern end of Larne Lough, Co. Antrim. *Ir. Nat. J.* **13**: 159-163.

LYNN, M. J. 1961b. Coastal Survey XI (New Series) Northern end of Larne Lough, Co. Antrim. *Ir. Nat. J.* **13**: 223-227.

LYNN, M. J. and McGURK, J. 1932. Botanical Society of Northern Ireland. Coastal Survey: VI. Ardglass - From the Pill to St. Patrick's Well. *Ir. Nat. J.* **4**: 114-117.

LYNN, M. J. and McGURK, J. 1934. Botanical Society of Northern Ireland. Coastal

Survey: IX. Killough and Coney Island. *Ir. Nat. J.* **5**: 52-56.

MAGGS, C. A. & GUIRY, M. D. 1985. Life history and reproduction of *Schmitzia hiscockiana* sp. nov. (Rhodophyta, Gigartinales) from the British Isles. *Phycologia* **24**: 297-310.

MAGGS, C. A. & GUIRY, M. D. 1987. *Gelidiella calcicola* sp. nov. (Rhodophyta) from the British Isles and Northern France. *Br. phycol. J.* **22**: 417-434.

MAGGS, C. A., DOUGLAS, S. E., FENETY, J. & BIRD, C. J. 1992. A molecular and morphological analysis of the *Gymnogongrus devoniensis* (Rhodophyta) complex in the North Atlantic. *J. Phycol.* **28**: 214-232.

MORTON, O. 1974. Marine algae of Sandeel Bay, Co Down. *Ir. Nat. J.* **18**: 32-35.

MORTON, O. 1978. Some interesting records of algae from Ireland. *Ir. Nat. J.* **19**: 240-242.

NORTON, T. A. 1970. The marine algae of County Wexford, Ireland. *Br. phycol. J.* **5**: 257-266.

PARKE, M. 1953. A preliminary check-list of British marine algae. *J. mar. biol. Ass. U.K.* **32**: 497-520.

PARKE, M. & DIXON, P. S. 1968. Check-list of British marine algae - second revision. *J. mar. biol. Ass. U.K.* **48**: 783-832.

SAWERS, W. 1854. List of algae gathered in the north of Ireland. *Naturalist, Morris,* **4**: 254-257.

SILVA, P. C. 1955. The dichotomous species of *Codium* in Britain. *J. mar. biol. Ass. U.K.* **34**: 565-577.

STANDEN, R. 1897. Some observations by English naturalists on the fauna of Rathlin Island and Ballycastle district. 1. General observations. *Ir. Nat.* **6**: 173-178.

THOMPSON, W. 1836. Abstract of a paper on Irish algae, read before the Natural History Society of Belfast on January 20. 1836. *Mag. Nat. Hist. London.* **9**: 147-151.

WADE, W. 1804. Plantae rariores in Hibernia inventae; or habitats of some plants, rather scarce and valuable, found in Ireland; with concise remarks on the properties and uses of many of them. *Trans. Dublin Soc.* **4**: 1-214.

WHITE, J. M. 1931. Botanical Society of Northern Ireland. Coastal Survey: III Warrenpoint to Rostrevor. *Ir. Nat. J.* **3**: 233-236.

WHITE, J. M. 1932. Botanical Society of Northern Ireland. Coastal Survey: IV. Rostrevor to Ballyedmond. *Ir. Nat. J.* **4**: 31-34.

WILCE, R.T. & MAGGS, C.A. 1989. Reinstatement of the genus *Haemescharia* (Rhodophyta, Haemeschariaceae fam. nov.) for *H. polygyna* and *H. hennedyi* comb. nov.(=*Petrocelis hennedyi*). *Can. J. Bot.* **67**: 1465-1479.

WRIGHT, W. S. & WRIGHT, J. W. 1933a. Botanical Survey of Northern Ireland. Coastal Survey: VII. Portrush. *Ir. Nat. J.* **4**: 196-198.

WRIGHT, W. S. & WRIGHT, J. W. 1933b. Botanical Survey of Northern Ireland. Coastal Survey: VIII. Dunluce. *Ir. Nat. J.* **4**: 241-243.

WYNNE, M.J. & MAGNE, F. 1991. Concerning the name *Fucus muscoides* (Cotton) J. Feldman et Magne. *Cryptogamie Algol.* **12**: 55-65.

Glossary

CRUSTOSE	Crustlike; said of thalli flattened against substrate.
CYSTOCARP	A reproductive structure of the Rhodophyta consisting of the diploid carposporophyte with any enclosing filaments.
DETERMINED	Confidently identified.
ECAD	A habitat form, showing characteristics imposed by the hatitat and non-heritable.
ENCRUSTING	See Crustose.
ENDOPHYTIC	Living within another plant but not parasitic.
EPIPHYTIC	Living on the surface of another plant but not parasitic.
EPIZOIC	Living on the surface of an animal.
EULITTORAL	The shore between the upper limit of the barnacle/fucoid zone and the upper limit of the laminarian zone. (Note 1).
FORMA	A taxonomic category below that of variety.
FROND	The erect part of a thallus.
GAMETOPHYTE	The phase of the algal life-history which bears gametes.
HAPLOID	Having a single set of chromosomes per nucleus.
HETEROMORPHIC	Having a life-history in which one phase differs morphologically from the other.
HYDROZOA	Marine animals of the class Coelenterata.
Incertae sedis	Of uncertain taxonomic position.
LITTORAL	The shore between the splash zone and the top of the laminarian zone. (Note 1).
MAERL	Loose-lying corallines - either living or dead.
MONOSIPHONOUS	Composed of a single row of cells.
PHENOTYPE	The sum of characteristics of a plant as opposed to the genotype.
RHODOLITH	Nodular forms of coralline algae growing on small free stones or shells etc.
SUBLITTORAL	Below the laminarian zone. (Note 1).
SUBSPECIES	The taxonomic devision of a species below the rank of species.
SYNONYM	One of two or more names applied to the same taxon.
TAXON (plural TAXA)	A taxonomic unit of any rank.
TETRASPORANGIAL	Having a sporangium of four spores.
THALLUS (plural THALLI)	The whole algal body.
UTRICLE	Inflated portion formed from one cell in some algae.
VARIETY	A taxonomic category below that of subspecies.

Note 1   The definitions of shore line noted here are based on Lewis, J. R. 1964. *The Ecology of Rocky Shores*. London. However the terms are used with slightly different meanings by different authors.

Index

All genera, species and infraspecific taxa in this account are included in this index.
Accepted taxa are shown in roman type; those detailed in this work and their principal page
number(s) are shown in **bold,** with synonyms in *italics.* English names are not included.